The Cambridge Manuals of Science and
Literature

SUBMERGED FORESTS

Buried Forest seen at low water at Dove Point, on the Cheshire coast

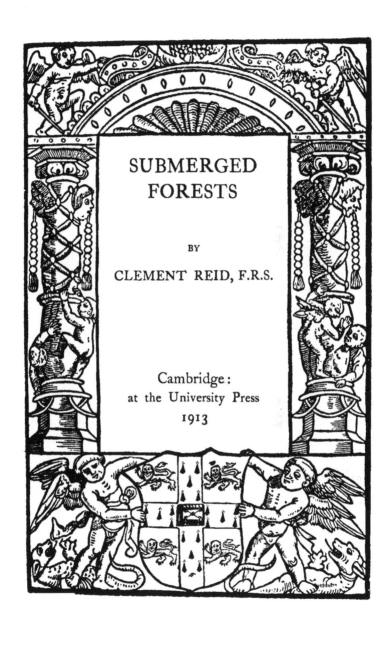

SUBMERGED FORESTS

BY

CLEMENT REID, F.R.S.

Cambridge:
at the University Press
1913

CAMBRIDGE UNIVERSITY PRESS
Cambridge, New York, Melbourne, Madrid, Cape Town,
Singapore, São Paulo, Delhi, Tokyo, Mexico City

Cambridge University Press
The Edinburgh Building, Cambridge CB2 8RU, UK

Published in the United States of America by
Cambridge University Press, New York

www.cambridge.org
Information on this title: www.cambridge.org/9781107401785

First published 1913
First paperback edition 2011

A catalogue record for this publication is available from the British Library

ISBN 978-1-107-40178-5 Paperback

*With the exception of the coat of arms
at the foot, the design on the title page is a
reproduction of one used by the earliest known
Cambridge printer, John Siberch, 1521*

PREFACE

KNOWLEDGE cannot be divided into compartments, each given a definite name and allotted to a different student. There are, and always must be, branches of knowledge in which several sciences meet or have an interest, and these are somewhat liable to be neglected. If the following pages arouse an interest in one of the by-ways of science their purpose has been fulfilled.

C. R.

February 17, 1913.

CONTENTS

LIST OF ILLUSTRATIONS

CHAPTER I

Most of our sea-side places of resort lie at the mouths of small valleys, which originally gave the fishermen easy access to the shore, and later on provided fairly level sites for building. At such places the fishermen will tell you of black peaty earth, with hazel-nuts, and often with tree-stumps still rooted in the soil, seen between tide-marks when the overlying sea-sand has been cleared away by some storm or unusually persistent wind. If one is fortunate enough to be on the spot when such a patch is uncovered this "submerged forest" is found to extend right down to the level of the lowest tides. The trees are often well-grown oaks, though more commonly they turn out to be merely brush-wood of hazel, sallow, and alder, mingled with other swamp-plants, such as the rhizomes of *Osmunda*.

These submerged forests or "Noah's Woods" as they are called locally, have attracted attention from early times, all the more so owing to the existence of an uneasy feeling that, though like most other

R.   1

geological phenomena they were popularly explained by Noah's deluge, it was difficult thus to account for trees rooted in their original soil, and yet now found well below the level of high tide.

It may be thought that these flats of black peaty soil though curious have no particular bearing on scientific questions. They show that certain plants and trees then lived in this country, as they do now ; and that certain animals now extinct in Britain once flourished here, for bones and teeth of wild-boar, wolf, bear, and beaver are often found. Beyond this, however, the submerged forests seem to be of little interest. They are particularly dirty to handle or walk upon ; so that the archaeologist is inclined to say that they belong to the province of geology, and the geologist remarks that they are too modern to be worth his attention ; and both pass on.

Should we conquer our natural repugnance for such soft and messy deposits, and examine more closely into these submerged forests, they turn out to be full of interest. It is largely their extremely inconvenient position, always either wet or submerged, that has made them so little studied. It is necessary to get at things more satisfactorily than can be done by kneeling down on a wet muddy foreshore, with the feeling that one may be caught at any time by the advancing tide, if the study is allowed to become too engrossing. But before leaving for a time the old land-surface exposed between tide marks, it will be well to note that we have already gained one piece of valuable information from this hasty traverse. We have learnt that the relative level of land and sea has changed somewhat, even since this geologically modern deposit was formed.

Geologists, however, sometimes speak of the submerged forests as owing their present position to various accidental causes. Landslips, compression of the underlying strata, or the removal of some protecting shingle-beach or chain of sand-dunes are all called into play, in order to avoid the conclusion that the sea-level has in truth changed so recently. The causes above mentioned have undoubtedly all of them affected certain localities, and it behoves us to be extremely careful not to be misled. Landslips cannot happen without causing some disturbance, and a careful examination commonly shows no sign of disturbance, the roots descending unbroken into the rock below. It is also evident in most cases that no landslip is possible, for the "forest" occupies a large area and lies nearly level.

Compression of the underlying strata, and consequent sinking of the land-surface above, is however a more difficult matter to deal with. Such compression undoubtedly takes place, and some of the appearances of subsidence since the Roman invasion are really cases of this sort. Where the trees of the

1—2

submerged forest can be seen rooted into hard rock, or into firm undisturbed strata of ancient date, there can, however, be no question that their position below sea-level is due to subsidence of the land or to a rise of the sea, and not to compression. But in certain cases it is found that our submerged land-surface rests on a considerable thickness of soft alluvial strata, consisting of alternate beds of silt and vegetable matter. Here it is perfectly obvious that in course of time the vegetable matter will decay, and the silt will pack more closely, thus causing the land-surface above slowly to sink. Subsidence of this character is well known in the Fenland and in Holland, and we must be careful not to be misled by it into thinking that a change of sea-level has happened within the last few centuries. The sinking of the Fenland due to this cause amounts to several feet.

The third cause of uncertainty above mentioned, destruction of some bank which formerly protected the forest, needs a few words. It is a real difficulty in some cases, and is very liable to mislead the archaeologist. We shall see, however, that it can apply only to a very limited range of level.

Extensive areas of marsh or meadow, protected by a high shingle-beach or chain of sand-dunes, are not uncommon, especially along our eastern coast. These marshes may be quite fresh, and even have trees growing on them, below the level of high tide, as long as the barrier remains unbroken. The reason of this is obvious. The rise and fall of the tide allows sea water to percolate landward and the fresh water to percolate seaward ; but the friction is so great as to obliterate most of the tidal wave. Thus the sea at high tide is kept out, the fresh water behind the barrier remaining at a level slightly above that of mean tide, and just above that level we may find a wet soil on which trees can grow. But, and here is the important point, a protected land-surface behind such a barrier can never lie below the level of mean tide ; if it sinks below that level it must immediately be flooded, either by fresh water or by sea water. This rule applies everywhere, except to countries where evaporation exceeds precipitation ; only in such countries, Palestine for instance, can one find sunk or Dead Sea depressions below mean-tide level of the open sea.

The submerged forest that we have already examined stretched far below the level of mean tide, in fact we followed it down to the level of the lowest spring tides. Nothing but a change of sea-level will account for its present position. In short, the three objections above referred to, while teaching us to be careful to examine the evidence in doubtful cases, cannot be accepted as any explanation of the constant and widespread occurrence of ancient land-surfaces passing beneath the sea.

We have thus traced the submerged forest down to low-water mark, and have seen it pass out of our reach below the sea. We naturally ask next, What happens at still lower levels? It is usually difficult to examine deposits below the sea-level; but fortunately most of our docks are excavated just in such places as those in which the submerged forests are likely to occur. Docks are usually placed in the wide, open, estuaries, and it is often necessary nowadays to carry the excavations fully fifty feet below the marsh-level. Such excavations should be carefully watched, for they throw a flood of light on the deposits we wish to examine.

Every dock excavation, however, does not necessarily cut through the submerged forests, for channels in an estuary are constantly shifting, and many of our docks happen to be so placed as to coincide with comparatively modern silted-up channels. Thus at King's Lynn they hit on an old and forgotten channel of the Ouse, and the bottom of the dock showed a layer of ancient shoes, mediaeval pottery, and suchlike—interesting to the archaeologist, but not what we are now in search of. At Devonport also the recent dock extension coincided with a modern silted-up channel. In various other cases, however, the excavations have cut through a most curious alternation of deposits, though the details vary from place to place.

The diagram (fig. 1) shows roughly what is found.
We will suppose that the docks are placed, as is
usually the case, on the salt marshes, but with their
landward edge reaching the more solid rising ground,
on which the warehouses, etc., are to be built.   Beginning at or just above the level of ordinary high-
water of spring tides, the first deposit to be cut
through is commonly a bed (A) of estuarine silt or
warp with remains of cockles, *Scrobicularia*, and
salt-marsh vegetation.   Mingled with these we find

Fig. 1.

drifted wreckage, sunk boats, and miscellaneous
rubbish, all belonging to the historic period.   The
deposits suggest no change of sea-level, and are
merely the accumulated mud which has gradually
blocked and silted up great part of our estuaries and
harbours during the last 3500 years.

This estuarine silt may continue downward to a
level below mean tide, or perhaps even to low-water
level ; but if the sequence is complete we notice
below it a sudden change to a black peaty soil (B),

full of vegetable matter, showing sallows, alder, and hazel rooted in their position of growth. In this soil we may also find seams of shell-marl, or chara-marl, such as would form in shallow pools or channels in a freshwater marsh. This black peaty soil is obviously the same "submerged forest" that we have already examined on the foreshore at the mouth of the estuary; the only difference being that in the more exposed situation the waves of the sea have cleared away the overlying silt, thus laying bare the land surface beneath. In the dock excavations, therefore, the submerged forest can be seen in section and examined at leisure.

The next deposit (*C*), lying beneath the submerged forest, is commonly another bed of estuarine silt, extending to a depth of several feet and carrying our observations well below the level of low-water. Then comes a second land-surface (*D*), perhaps with trees differing from those of the one above; or it may be a thick layer of marsh peat. More silt (*E*) follows; another submerged forest (*F*); then more estuarine deposits (*G*); and finally at the base of the channel, fully 50 feet below the level of high-water, we may find stools of oak (*H*) still rooted in the undisturbed rock below.

As each of these deposits commonly extends continuously across the dock, except where it happens to abut against the rising ground, it is obvious that it is absolutely cut off from each of the others. The lowest land-surface is covered by laminated silts, and that again is sealed up by the matted vegetation of the next growth. Thus nothing can work its way down from layer to layer, unless it be a pile forcibly driven down by repeated blows. Materials from the older deposits in other parts of the estuary may occasionally be scoured out and re-deposited in a newer layer; but no object of a later period will find its way into older beds.

Thus we have in these strongly marked alternations of peat and warp an ideal series of deposits for the study of successive stages. In them the geologist should be able to study ancient changes of sea-level, under such favourable conditions as to leave no doubt as to the reality and exact amount of these changes. The antiquary should find the remains of ancient races of man, sealed up with his weapons and tools. Here he will be troubled by no complications from rifled tombs, burials in older graves, false inscriptions, or accidental mixture. He ought here to find also implements of wood, basket-work, or objects in leather, such as are so rarely preserved in deposits above the water-level, except in a very dry country.

To the zoologist and botanist the study of each successive layer should yield evidence of the gradual changes and fluctuations in our fauna and flora, during early periods when man, except as hunter, had little influence on the face of nature. If I can persuade observers to pay more attention to these modern deposits my object is secured, and we shall soon know more about some very obscure branches of geology and archaeology.

I do not wish to imply that excellent work has not already been done in the examination of these deposits. Much has been done; but it has usually been done unsystematically, or else from the point of view of the geologist alone. What is wanted is something more than this—the deposits should be examined bed by bed, and nothing should be overlooked, whether it belong to geology, archaeology, or natural history. We desire to know not merely what was the sea-level at each successive stage, but what were the climatic conditions. We must enquire also what the fauna and flora were like, what race of man then inhabited the country, how he lived, what weapons and boats he used, and how he and all these animals and plants were able to cross to this country after the passing away of the cold of the Glacial period.

To certain of the above questions we can already make some answer; but before dealing with conclusions, it will be advisable to give some account of the submerged land-surfaces known in various parts of Britain. This we will do in the next chapters.

Before going further it will be well to explain and limit more definitely the field of our present enquiry. It may be said that there are "submerged forests" of various geological dates, and this is perfectly true. The "dirt-bed" of the Isle of Purbeck, with its upright cycad-stems, was at one time a true submerged forest, for it is overlain by various marine strata, and during the succeeding Cretaceous period it was probably submerged thousands of feet. Every coal seam with its underlying soil or "underclay" penetrated by stigmarian roots was also once a submerged forest. Usage, however, limits the term to the more recent strata of this nature, and to these we will for the present confine our attention. We do not undertake a description of the earlier Cromer Forest-bed, or even of the Pleistocene submerged forests containing bones of elephant and rhinoceros and shells of *Corbicula fluminalis*. These deposits will, however, be referred to where from their position they are liable to be confounded with others of later date.

CHAPTER II

THE THAMES VALLEY

IN the last chapter an attempt was made to give a general idea of the nature of the deposits; we will now give actual examples of what has been seen.

Unfortunately we cannot say "what can be seen," for the lower submerged forests are only visible in dock excavations. As these works are carried well below the sea-level and have to be kept dry by pumping, it is impossible for them to remain open long, and though new excavations are constantly being made, the old ones are nearly always hidden within a few weeks of their becoming visible. Of course these remarks do not apply to the highest of these submerged land-surfaces, which can be examined again and again between tide-marks, whenever the tide is favourable and the sand of the foreshore has been swept away.

The most convenient way of dealing with the evidence will perhaps be to describe first what has been seen in the estuary of the Thames. Then in later chapters we will take the localities on our east coast and connected with the North Sea basin. Next we will speak of those on the Irish Sea and English Channel. Lastly, the numerous exposures on the west or Atlantic coast will require notice, and with them may be taken the corresponding deposits on the French coast. Each of these groups will require a separate chapter.

The Thames near London forms a convenient starting point, for the numerous dock-excavations, tunnels, deep drains and dredgings have laid open the structure of this valley and its deposits in an exceptionally complete way. The published accounts of the excavations in the Thames Valley are so voluminous that it is impossible here to deal with them in any detail; we must therefore confine ourselves to those which best illustrate the points we have in view, choosing modern excavations which have been carefully watched, noted, and collected from rather than ancient ones.

We cannot do better than take as an illustration of the mode of occurrence and levels of the submerged land-surfaces the section seen in the excavation of Tilbury Docks, for this was most carefully noted by the engineers, and was visited by two competent observers, Messrs. W. Whitaker and F. C. J. Spurrell. This excavation is of great scientific importance, for it led to the discovery of a human skeleton beneath three distinct layers of submerged peat, and these remains have been most carefully studied by Owen and Huxley, and more recently by Professor Keith.

The section communicated to Sir Richard Owen by Mr Donald Baynes, the engineer superintending the excavation at the time of the discovery, is shown in the diagram on p. 14. As Mr Baynes himself saw part of the skeleton in the deposit, his measured section is most important as showing its exact relation to the submerged forests. It is also well supplemented by the careful study of the different

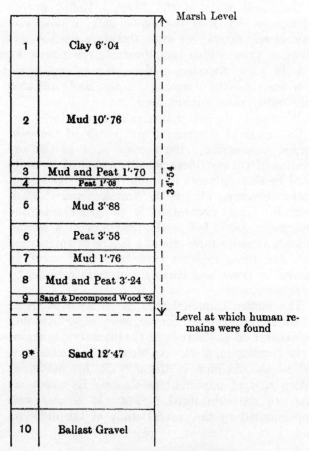

Fig. 2. Section at Tilbury Docks.

layers made by Mr Spurrell, for though his specimens did not come from exactly this part of the docks, the various beds are traceable over so large an area that there is no doubt as to their continuity.

Owen thought that this skeleton belonged to a man of the Palaeolithic period, considering it contemporaneous with the mammoth and rhinoceros found elsewhere in the neighbourhood. Other geological writers showed however that these deposits were much more modern, and some of them spoke somewhat contemptuously of their extremely recent date. But Huxley saw the importance of this "river-drift man" as an ancient and peculiar race, and Professor Keith has more recently drawn especial attention to the well-marked characteristics of the type . The skeleton is not of Palaeolithic date, but neither is it truly modern; other examples have turned up in similar deposits elsewhere.

We will now describe more fully the successive layers met with in Tilbury Docks, condensing the account from that given by Messrs. Spurrell and Whitaker, and using where possible the numbers attached by the engineer to the successive beds.

It will be noticed that the marsh-level lies several feet below Trinity high water. Below the sod of the marsh came a bed of fine grey tidal clay (1), in which at a depth of seven feet below the surface Mr Spurrell noted, in one part of the docks, an old grass-grown surface strewn with Roman refuse, such as tiles, pottery, and oyster shells. This fixes the date of the layer above as post-Roman; but the low position of the Roman land-surface, now at about mean-tide level, is due in great part to shrinkage since the marsh was embanked and drained—it is unconnected with any general post-Roman subsidence of the land.

Beneath the Roman layer occurs more marsh-clay and silt (2), resting on a thin peat (4) which according to Mr Whitaker is sometimes absent. Then follows another bed of marsh-clay (5), shown by the engineer as four feet thick, but which in places thickens to six or seven feet. Below this is a thick mass (6) of reedy peat (the "main peat" of Mr Spurrell), which is described as consisting mainly of *Phragmites* and *Sparganium*, with layers of moss and fronds of fern. The other plants observed in this peat were the elder, white-birch, alder and oak. Associated with them were found several species of freshwater snails and a few land forms; but the only animal or plant showing any trace of the influence of salt water was *Hydrobia ventrosa*, a shell that requires slightly brackish water.

The main peat rests on another bed of estuarine silt (7 and 8), which seems to vary considerably in thickness, from 5 to 12 feet. It is not quite clear from the descriptions whether the "thin woody peat" of Mr Whitaker and the "sand with decayed wood"

(9) of the engineer represent a true growth in place, like the main peat; it is somewhat irregular and tends to abut against banks of sand (9*) rising from below. In one of these banks, according to Mr Spurrell, the human skeleton was found.

The contents of the sand (other than the skeleton) included *Bythinia* and *Succinea*; and as Mr Spurrell calls it a "river deposit," it apparently did not yield estuarine shells, like the silts above. The subangular flint gravel (10) below has all the appearance of a river gravel; it may be from 10 to 20 feet thick, and rests on chalk only reached in borings.

The floor of chalk beneath these alluvial deposits lies about 60 or 70 feet below the Ordnance datum in the neighbourhood of Tilbury and Gravesend, and in the middle of the ancient channel of the Thames it may be 10 feet lower; but there is no evidence of a greater depth than this. We may take it therefore that here the Thames once cut a channel about 60 feet below its modern bed. We cannot say, however, from this evidence alone that the sea-level then was only 60 feet below Ordnance datum, for it is obvious that it may have been considerably lower. If, as we believe, the southern part of the North Sea was then a wide marsh, the Thames may have followed a winding course of many miles before reaching the sea, then probably far away, in the latitude of the Dogger Bank. This must be borne in mind: we

R.   2

know the minimum extent of the change of level; but its full amount has to be ascertained from other localities.

This difficulty has seemed of far greater importance than it really is, and some geologists have suggested that at this period of maximum elevation, England stood several hundred feet higher above the sea than it does now. I doubt if such can have been the case. Granting that the sea may have been some 300 miles away from Tilbury, measured along the course of the winding river, this 300 miles would need a very small fall per mile, probably not more than an inch or two. The Thames was rapidly growing in volume, from the access of tributaries, and was therefore flowing in a deeper and wider channel, which was cut through soft alluvial strata; it therefore required less and less fall per mile. Long before it reached the Dogger it probably flowed into the Rhine, then containing an enormous volume of water and draining twice its present catchment area.

The clean gravel and sand which occupy the lower part of the ancient channel at Tilbury require to be more closely examined, for it is not clear that they are, as supposed, of fluviatile origin; they may quite well be estuarine. In the sand Mr Spurrell found the freshwater shells *Bythinia* and *Succinea*, and in it was also found the human skeleton described by Owen; but, according to Mr Spurrell, on the surface of this sand lay a few stray valves of the estuarine *Scrobicularia* and of *Tellina*. The bottom deposits were probably laid down in a tidal river; but whether within the influence of the salt water is doubtful.

As far as the Tilbury evidence goes it suggests a maximum elevation of the land of about 80 feet above its present level; but we will return to this question when we have dealt with the other rivers flowing more directly into deep sea. The animals and plants found at Tilbury were all living species.

It is unnecessary here to discuss more fully the submerged forests seen in dock and other excavations in the Thames flats, for they occupy a good many pages in the *Geology of London* published by the Geological Survey. Even 250 years ago, the hazel trees were noticed by the inquisitive Pepys during one of his official visits to the dockyards, and later writers are full of remarks on the ancient yew trees and oaks found well below the sea-level. Most of these early accounts are, however, of little scientific value.

CHAPTER III

THE EAST COAST

It is not our purpose to describe in detail the many exposures of submerged land-surfaces which have been seen on the shores of the North Sea. This

would serve no useful purpose and would be merely
tedious. We need only say that the floor of Eocene or
Cretaceous strata on which these ancient subaerial
deposits rest is constantly found at depths of 50 or
60 feet below the level of the existing salt-marsh.
But where, as in the estuary of the Thames and
Humber, an older channel underlies a modern
channel, the floor sinks about 30 feet lower. From
present marsh-level to ancient marsh-level is about
60 feet; from present river-bottom to old river-
bottom is also about 60 feet. This, therefore, is the
extent of the former elevation, unless we can prove
that the sea was then so far away that the river once
had many miles to flow before reaching it. This is
the point we have now to consider as we trace the
submerged forests northward and towards the deeper
seas.

Before we leave the southern part of the North
Sea basin it will be well to draw attention to a few
of the half-tide exposures which for one reason or
another may tend to mislead the observer. The mere
occurrence of roots below tide marks is not sufficient
to prove that the land-surfaces seen are all of one date.

Not far from Tilbury is found the well-known
geological hunting ground of Grays, where the brick-
yards have yielded numerous extinct mammalia and
several land and freshwater shells now extinct in
Britain. These deposits lie in an old channel of the

Thames, cut to below mean-tide level, but here not coinciding exactly in position either with the channel of the existing river, or with the channel in which the submerged forests lie.

It is fortunate that the channels do not coincide, for this enables us to distinguish the more ancient deposits. A glance at a geological map shows, however, that they must coincide elsewhere, and where the Thames has re-occupied its old channel it is clear that the destruction of the earlier deposits may have led to a mixture of fossils and implements belonging to three different dates. Mammoth teeth and Palaeolithic implements, Irish elk and polished stone implements, may all be dredged up in the modern river gravel, associated with bits of iron chain, old shoes, and pottery. Such a mixture does actually occur in the Thames estuary, and it makes us hesitate to accept the teeth of mammoth which were dredged in the Thames as really belonging to so late a period as that of the submerged forests.

At Clacton a similar difficulty is met with, for there again an ancient channel contains alternating estuarine and freshwater deposits with layers of peat, and is full of bones belonging to rhinoceros, hippopotamus, elephant and other extinct mammalia. Of course the peat-beds in this channel are just as much entitled to the name "submerged forest" as the more modern deposits to which recent usage restricts it.

They belong, however, to another and more ancient
chapter of the geological record than that with which
we are now dealing.   I do not say a less interesting
one, for they are of the greatest importance when we
study the times when Palaeolithic man flourished; but
at present we have as much as we can do to understand the later deposits and to realize the great
changes to which they point.   We must not turn
aside for everything of interest that we come across
in this study ; these earlier strata are worthy of a
book to themselves.

As we travel northward along the coast, again and
again we meet with evidence of a submerged nearly
level platform, "basal plane," or ancient "plane of
marine denudation," lying about 50 feet below the
sea.   We find it at Langer Fort, which lies opposite
to Harwich on a spit of sand and shingle stretching
across Harwich Harbour.   Here the floor of London
Clay was met with in a boring at 54 feet below the
surface.

The Suffolk coast north of Southwold yields yet
another complication, for between Southwold and
Sherringham in Norfolk there appears at the sea-
level a land-surface considerably more ancient than
anything we have yet been dealing with.   This is
the so-called "Cromer Forest-bed," which consists of
alternating freshwater and estuarine beds, with
ancient land-surfaces and masses of peat.   It contains

numerous extinct mammals, mainly of species older than and different from those of Clacton and Grays.

The mammalian remains differentiate these deposits at once ; but if no determinable mammals are found, the crushing of the bones and the greater compression and alteration of the peaty beds serves to distinguish them, for this Forest-bed dates back to Pliocene times, passes under a considerable thickness of glacial beds, and has been over-ridden by the ice-sheet during the Glacial epoch.

The Cromer Forest-bed has been exposed particularly well of late years at Kessingland, near Lowestoft, where the sea has encroached greatly. It is well worth while to make a comparative study of this deposit, of the Grays and Clacton *Cyrena*-bed, of the submerged forests of the Thames docks, and of the strata now being formed in and around the Norfolk Broads. By such a comparison we can trace the effects of similar conditions occurring again and again. The fauna and flora slowly change, species come and go, man appears and races change : though the same physical conditions may recur life ever changes.

The Norfolk Broads, just referred to, deserve study from another point of view : their origin is directly connected with the submergence which forms the theme of this book. These broads are shallow lakes, always occupying part of the widest alluvial flats which border the rivers ; but they are usually

out of the direct course of the present river ; they
therefore receive little of the sediment brought down
in flood-time. On the other hand they are steadily
being filled up with growing vegetation and turned
into peat mosses.

The origin of these shallow freshwater lakes, which
form a characteristic feature in the scenery of East
Anglia, has been much debated ; but with the knowledge obtained from a study of the submerged forests
the explanation is perfectly simple. During this
period of slow submergence each of the shallow
valleys in which the broads now lie was turned into
a wide and deep navigable estuary, which extended
inland for many miles. When the subsidence stopped
the sea and tides soon formed bars and sand-banks
at the mouths of the estuaries, and lateral tributaries
pushed their deltas across. The Norfolk rivers, being
small and sluggish, were driven to one side, and could
neither cut away the sand-banks nor fill up with
sediment such wide expanses. These estuaries therefore were silted up with tidal mud and turned into
irregular chains of lakes, separated by irregular
bars and sand-banks. The lakes, instead of becoming rapidly obliterated and filled up by deltas
which crept gradually seaward, remained as freshwater broads ; for as soon as a bank became high
enough for the growth of reeds and sedges the river
mud was strained out and only nearly clean water

reached the lagoon behind. Thus a depression once left, provided it was out of the direct course of the river, tended to remain as a freshwater lake until vegetable growth could fill it, and the river mud was spread out over the salt-marshes or went to raise the sand-banks till they became alluvial flats, and thus still more thoroughly isolated the broad.

A few centuries will see the disappearance of the last of the broads, which have silted up to an enormous extent within historic times ; but the fact that so many of these broads still exist may be taken as clear evidence of the recent date of the depression which led to their formation.

When we look at ancient records, and notice the rapidity with which the broads and navigable estuaries are becoming obliterated, we cannot help wondering whether the measure of this silting up may not give us the date of the last change of sea-level. It should do so if we could obtain accurate measurements of the amount of sediment deposited annually, of the rate at which the sea is now washing it in, and of the rate at which the rivers are bringing it down. All these factors, however, are uncertain, and it is particularly difficult to ascertain the part played by the muddy tidal stream which flows in after storms and spreads far and wide over the marsh.

Though all the factors are so uncertain, we can form some idea of the date of the submergence.

Many years ago I made a series of calculations, founded on the silting up of our east coast estuaries, the growth of the shingle-spits, and the accumulation of sand-dunes. The results were only roughly concordant, but they seemed to show that the subsidence stopped about 2500 years ago and was probably still in progress at a date 500 years earlier. This question of dates will be again referred to in a later chapter.

Before leaving the Broad district we must refer to a boring made at Yarmouth, which, according to Prof. Prestwich, showed that the recent estuarine deposits are there 120 feet thick, and consequently that the ancient valley was far deeper than any recorded in the foregoing pages. There is no doubt, however, that this interpretation is founded on a mistake, for other borings at Yarmouth, Lowestoft, and Beccles came to muddy sands and clays belonging to the upper part of the Crag, now known to thicken greatly eastward. The recent deposits descend only to a depth of about 50 feet at Yarmouth, and consist of sand and shingle ; the beds below contain Pliocene mollusca. This emendation is also borne out by the entirely different character of the recent estuarine deposits at Potter Heigham, where we again find a submerged forest at about 56 feet below the marsh-level. The section recorded by Mr Blake is as follows :—

	feet
Bluish-grey loam	24
Grey silty sand	½ to 2
Stiff bluish-grey loam, clay, and silt full of	
cockles, &c.	13
Black peat, hard, and much compressed ...	17
White and buff sand	2
	58

It will be noticed that here only one peat bed was found, and was at the usual depth of the lowest submerged forest. Possibly the white sand below was the bleached top of the Crag; but this point was not cleared up.

If we resume our journey northward along the Norfolk coast we come to the well-known locality of Eccles, where the old church tower described and figured by Lyell in his *Principles of Geology* long stood on the foreshore, washed by every spring tide. The position of this church formed a striking illustration of the protection afforded by a chain of sand-dunes. The church was originally built on the marshes inside these dunes, at a level just below that of high-water spring tides. But as the dunes were driven inland they gradually overwhelmed the church, till only the top of its tower appeared above the sand. In this state it was pictured by Lyell in the year 1839. Later on (in 1862) it was again sketched by the Rev. S. W. King, and stood on the seaward side of the dune and almost free from sand. For a series

of years, from 1877 onward, I watched the advance of
the sea, and as the church tower was more and more
often reached by the tides, its foundations were laid
bare and attacked by the waves, till at last the tower
fell.

Not only were the foundations of Eccles church
exposed on the foreshore, but an old road across the
marshes also appeared on the seaward side of the
dunes, giving a still more exact idea of the former
great influence of the chain of dunes in damping the
oscillations of the tidal wave. The tide outside now
rises and falls some 12 or 15 feet; on the marsh
within its influence is only felt under exceptional
circumstances. A road across the marsh at a level
four or five feet below high-water, as this one stood,
would still be passable, except during unusual floods.

Eccles Church is an excellent example of the way
in which an ancient land-surface may now be found
below the level of high-water, and yet no subsidence
of the land has taken place. But this coast can give
even more curious examples. It does not need a sand-
dune to deaden the rise and fall of the tides; even
a submerged bank will have much the same effect.
Extensive submerged sand-banks extend parallel
with the coast, protecting the anchorage known as
Yarmouth Roads. These banks rise so nearly to the
surface of the sea that not only do they protect the
town and anchorage against the waves, they deaden

the tidal oscillation to such an extent that its range
is much greater outside the bank than within.

If these submerged outer banks were to be swept
away by some change in the set of the currents, large
areas now cultivated and inhabited would be flooded
by salt water at every spring tide, and the turf of the
meadows would be covered by a layer of marine silt,
such as we see alternating with the submerged forests
in the docks of the Thames.   Such alternations, if
thin, do not necessarily prove a change in the level
of the sea; they may only point to the alternate
accumulation and removal of sand-banks in a distant
part of the estuary.

The Norfolk coast trends westward soon after
leaving Cromer, and where the cliff seems to pass
inland at Weybourn we enter an ancient valley, one
side of which has been entirely cut away by the sea,
except for a few relics of the further bank, now
included in the shingle beach which runs out to sea
nearly parallel with the coast and protects Blakeney
Harbour.   Here again we find that in the bottom of
the valley there must be a submerged forest, for
slabs of peat are often thrown up at Weybourn, and
by the use of a grapnel the peat was found in place
off Weybourn at a depth of several fathoms.

When the coast turns southward again, and the
wide bay of the Wash is entered, we find an extensive development of submerged land-surfaces and peat

beds, extending over great part of the Fenland. In
fact the whole Fenland and Wash was once a slightly
undulating plain, cut into by numerous shallow open
valleys. The effect of the submergence of this area
has been to cause the greater part of it to silt up to
a uniform level, through the accumulation of warp
and growth of peat; so that now the Fenland has
become a dead level, out of which a few low hills rise
abruptly. The islands of the Fenland, such as those
on which Ely and March are built, are merely almost
submerged hill-tops; they were not isolated by marine
action.

It is obvious that a wide sheltered bay of this sort
forms an ideal area in which to study the gradual
filling up and obliteration of the valleys, as the land
sank; and it may enable us to learn the maximum
amount of the change of sea-level. The Fenland
unfortunately does not contain very deep dock
excavations, and we have only various shallower
engineering works to depend on, though numerous
borings reach the old floor.

A preliminary difficulty, however, meets us in the
study of the Fen deposits; it is the same difficulty
that we have already referred to when describing
Clacton and Grays, and we shall meet with it again.
In certain parts of the Fenland, particularly about
March and Chatteris, a sheet of shoal-water marine
gravelly sand caps some of the low hills, which rise a

few feet above the fen-level. The gravel for long was taken to be the same bed that passes under the marshes. Later work showed however that these gravels, with their sub-arctic marine fauna and containing also *Corbicula fluminalis*, were of much earlier date than the true fen-deposits. Just as we saw happen in the Thames Valley, a wide plain and estuary existed long before the deeper channels containing the submerged forests were cut; and the deposits of this older estuary and its tributaries are still to be found in patches here and there. Sometimes, as at March, they cap hills a few feet above the fen-level; but as often they fill channels not quite coinciding with the later channels; just as they do at Grays. Or two deposits of quite different date may lie side by side, as they do in the Nar Valley, or at Clacton, or on the Sussex coast.

The true fen-deposits were carefully examined by Messrs. Marshall, Fisher, and Skertchly, as far as the shallow sections would allow, and the following account is mainly condensed from that given by Mr Skertchly in his *Geology of the Fenland.*

During the excavation of certain deep dykes for the purpose of draining the fens there was discovered at a depth of about 10 feet below the surface a forest of oaks, with their roots imbedded in the underlying Kimmeridge Clay. The trunks were broken off at a height of about three feet. Some of the fallen trees

were of fine proportion, measuring three feet in
diameter, quite straight and seldom forked. At an
average height of two feet above this "forest No. 1"
the remains of another were found (in the peat)
consisting of oaks and yews. Three feet above
"forest No. 2" lay the remains of another, in which
the trees are all Scotch firs, some of which were three
feet in diameter. Above this and near to the surface
was seen a still newer forest of small firs. The peat
close to the surface contains remains of sallow and
alder, and was formed with the sea at its present
level.

It will be noticed that the greatest depth at which
these rooted trees were found was only about ten feet
below the sea-level. At this high level we must
expect to find that the growth of the peat was
practically continuous, and that the different submerged forests run together. In adjoining depressions
the different forests would occur at lower levels and
would be separated by beds of marine silt. It does
not follow from the position that a low-level submerged land-surface is older than one at a higher
elevation, for above the present sea-level all these
stages are represented by a few inches of soil, on
which forest after forest has grown and decayed.
Anyone who has collected antiquities on fields knows
what a curious jumble of Palaeolithic, Neolithic,
bronze age, Roman, mediaeval and recent things may

be found mixed in these few inches of soil, or may be thrown up by an uprooted tree. The great advantage of studying the deeply submerged forests is that in them the successive stages are separated and isolated, instead of being mingled in so confusing a fashion.

For further information as to the more deeply submerged land-surfaces we may turn to the numerous records of borings made in the Fenland and collected by the Geological Survey. These show that the thickness of the fen-deposits varies considerably from place to place, that the floor below undulates and is by no means so flat as the surface of the fen above. Most of these borings, however, were not continued through the gravels which lie at the base of the deposit, and thus we can only be certain of the total depth to the Jurassic clay or boulder clay in a few places. The maximum thickness of the fen-beds yet penetrated is less than 60 feet, and a submerged forest was found at Eaubrink at about 40 feet. It is possible however that none of these scattered borings has happened to hit upon one of the buried river-channels, which formerly wandered through this clayey lowland; if one were found it would probably show that the alluvial deposits are somewhat thicker than these measurements, and that they descend to a depth about equal to that reached in the valleys of the Thames or Humber.

It is useless to discuss in more detail the lower

R. 3

submerged forests of the Fenland, for we cannot get at them to examine them properly. They have been as effectually overwhelmed and hidden as the remains of King John's baggage train, which has never been seen again since it wandered off the flooded causeway during the disastrous spring-tide of October 11, 1216, and sank into the soft clay and quicksands.

The higher submerged forests of the Fenland are however of great interest, and as already pointed out they have been exposed to view in cutting the fen dykes, especially near Ely. Perhaps a closer study of these might enable us to arrive at some idea of the time taken for the growth of a series of forests of this sort, and for the accompanying mass of peat. The variations in the flora also need more exact analysis before we can say what they betoken. The oak-forest at the bottom is what we should expect on a clay soil; but the reason for the succession of trees above is not obvious. It need not necessarily point to climatic change, though it may do so; but it certainly looks as if the peaty bogs were alternately wetter and drier, so that sometimes moss grew, and sometimes fir-trees. Neither need this change imply an up-and-down movement of the land, though it may be due to such a cause.

Subsidence would destroy the oaks and allow a peat-moss to form; but if the subsidence were intermittent the moss would increase in thickness, become

more compact, and its surface rise, till it was dry
enough for pines. Another subsidence would cause
spongy peat again to spread and kill the pine, and
so on. Intermittent subsidence seems sufficient to
account for all the changes of vegetation we have
yet noticed in connexion with these submerged land-
surfaces.

Of the fauna of the fen-silts and peats it is very
difficult to give any satisfactory account. If we put
aside the March and Chatteris marine gravels with
Corbicula fluminalis, and the Nar Valley Clay with
its northern marine mollusca as being of older date ;
and if we also reject the marginal gravels with
hippopotamus and mammoth as being more ancient,
there only remain a few mammals such as the beaver,
wolf, wild boar, and certain cetacea, which we can be
sure came out of the true fen-deposits. Implements
made by man have only been found in the higher
layers, and there seems to be no record in this area
of a stone implement found below a submerged
forest.

Submerged forests of the ordinary type are often
to be seen between tide-marks on the flat shores of
Lincolnshire ; but as they still await proper study
they need not here detain us, and we will pass on
to the next large indentation of the coast-line, the
estuary of the Humber.

Here, owing to the excavation of extensive docks,

and to a series of trial borings for a tunnel beneath
the Humber, the structure of the valley has been
clearly laid open. It is much the same as that of the
Thames ; but as we are in a glaciated area we find,
as in the Fenland, that much of the erosion had taken
place before or during the Glacial Epoch, for boulder
clay occupies part of the valley.

Boulder clay or till not only occupies part of the
valley, it descends far below the present river bottom
and even below the lowest submerged forest. This
we find always to be the case in the glaciated parts

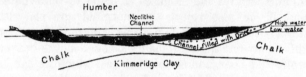

Fig. 3.

of Britain ; but whether the deep trenching is due to
the ploughing out of a trough by a tongue of the ice-
sheet, to sub-glacial streams below sea-level, or to
erosion by a true sub-aerial river is still a doubtful
point. However, this question must not detain us ;
we are not now dealing with elevations and depressions of so ancient a date, and must confine our
attention to post-glacial movements.

The section shown in fig. 3 will explain better
than any words the structure of the Humber Valley.

It is drawn to scale from the engineer's section, and
shows at a glance the three channels. The deepest
and widest channel is that occupied by glacial
deposits; an intermediate channel (shown in black)
is occupied by silt and submerged forests; and a
shallower channel is occupied by the present Humber
and its alluvium. One interesting point, however,
this section does not happen to illustrate. Somewhat
lower down the Humber we come to gravels and silts
full of sub-arctic marine mollusca and *Corbicula
fluminalis*, exactly as in regions further south, and
presumably of the same age as the deposits we have
already mentioned as found at March in the Fenland
and at Grays in Essex. The exact relation of these
Corbicula-beds to the deep channel filled with glacial
drift, below the marshes of the present Humber, is
still somewhat uncertain, but the marine beds clearly
rest on boulder clay, and seem also to be overlain by
another glacial deposit.

The section leaves no doubt that in post-glacial
times the Humber cut a channel about 60 feet below
its present bed, or to just the same depth as did the
Thames. This may possibly be an accidental coincidence; but it is very suggestive that both these rivers
should have cut their beds to the same depth. Such
coincidences suggest that we are dealing with a period
when each of our great rivers was able to cut to a
definite base-level, below which it could not go. This

base-level must either have been the sea, or some
vast alluvial plain then occupying the bed of the
North Sea. In either case the plain must then have
been fully 60 feet lower than the present sea-level.
Not only did the ancient Humber cut to the same
depth as the ancient Thames, but in each area the
ancient river was flanked by a wide alluvial flat which
now lies from 40 to 60 feet below the modern marsh
level.

The flat coast of Holderness, which stretches from
the Humber northward to Flamborough Head, shows
also occasional submerged forests; but the want of
excavations beneath the sea-level makes it impossible
to say much about them. North of Flamborough
Head it seems as though depression gave place to
elevation, and when we pass into Scotland the Neolithic deposits seem to be raised beaches instead of
submerged forests. We need not therefore devote
more time to a consideration of the details connected
with the submerged land-surfaces which border the
lands facing the North Sea. They evidently once
formed part of a wide alluvial flat stretching seaward
and running up all our larger valleys. We must
now consider how far seaward this plain formerly
extended.

Here, fortunately, we meet with a most surprising
piece of evidence, which adds enormously to the importance of this plain, and shows that the submergence

is no local phenomenon, but a widespread movement of depression which must greatly have altered the physical geography of north-western Europe during times within the memory of man. This evidence deserves a separate chapter.

CHAPTER IV

THE DOGGER BANK

For the last 50 years it has been known to geologists that the bed of the North Sea yields numerous bones of large land animals, belonging in great part to extinct species. These were first obtained by oyster-dredgers, and later by trawlers. Fortunately a good collection of them was secured by the British Museum, where it has been carefully studied by William Davies. The bones came from two localities. One of them, close to the Norfolk coast off Happisburgh, yielded mainly teeth of *Elephas meridionalis*, and its fossils were evidently derived from the Pliocene Cromer Forest-bed, which in that neighbourhood is rapidly being destroyed by the sea. This need not now detain us.

The other locality is far more extraordinary. In the middle of the North Sea lies the extensive shoal known as the Dogger Bank, about 60 or 70 miles

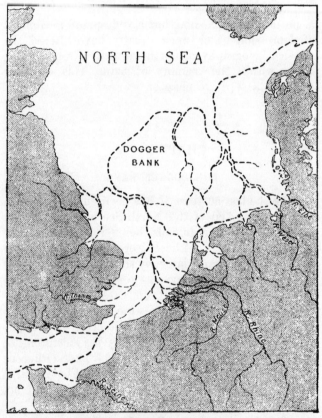

Fig. 4.—Showing approximate Coast-line at the period of
the lowest Submerged Forest.

from the nearest land. This shoal forms a wide irregular plateau having an area nearly as big as Denmark. Over it for the most part the sea has a depth of only 50 or 60 feet; all round its edge it slopes somewhat abruptly into deeper water, about 150 feet in the south, east, and west, but much deeper on the north. This peculiar bank has been explained as an eastward submerged continuation of the Oolite escarpment of Yorkshire; or, alternatively, as a mere shoal accumulated through the effects of some tidal eddy; but neither of these explanations will hold, for Oolitic rocks do not occur there, and the bank has a core quite unlike the sand of the North Sea.

When trawlers first visited the Dogger Bank its surface seems to have been strewn with large bones of land animals and loose masses of peat, known to the fishermen as "moorlog," and there were also many erratic blocks in the neighbourhood. As all this refuse did much damage to the trawls, and bruised the fish, the erratics and bones were thrown into deeper water, and the large cakes of moorlog were broken in pieces. A few of the erratics and some of the bones were however brought to Yarmouth as curiosities. Now the whole surface of the Dogger Bank has been gone over again and again by the trawlers, and very few of the fossil bones are found; unfortunately no record seems to have been kept as to the exact place where these bones were trawled.

The species found were :—

Ursus (bear)	Bison priscus (bison)
Canis lupus (wolf)	Equus caballus (horse)
Hyaena spelaea (hyaena)	Rhinoceros tichorhinus (woolly
Cervus megaceros (Irish elk)	rhinoceros)
,, tarandus (reindeer)	Elephas primigenius (mammoth)
,, elaphus (red-deer)	Castor fiber (beaver)
,, a fourth species	Trichechus rosmarus (walrus)
Bos primigenius (wild ox)	

Though mammalian bones are now so seldom found, whenever the sand-banks shift slightly, as they tend to do under the influence of tides and currents, the edges of the submerged plateau are laid bare, exposing submarine ledges of moorlog, which still yield a continuous supply of this material. Messrs. Whitehead and Goodchild have recently published an excellent account of it, having obtained from the trawlers numerous slabs of the peculiar peaty deposit, with particulars as to the latitude and longitude in which the specimens were dredged. Mrs Reid and I have to thank the authors for an opportunity of examining samples of the material, which has yielded most interesting evidence as to the physical history, botany, and climatic conditions of this sunken land. The following account is mainly taken from their paper and our appendix to it.

We are still without information as to the exact positions of the submarine ledges and cliffs of peat from which the masses have been torn; but there seems little doubt that some of them were actually torn off by the trawl. One block sent to me was full of recently dead half-grown *Pholas parva*, all of one age, and must evidently have been torn off the solid ledge. *Pholas* never makes its home in loose blocks. We unfortunately know very little about the natural history of the boring mollusca and their length of life. If, as I think, this species takes two years to reach full growth, then it is evident that the ledge of moor-log full of half-grown specimens must have been exposed to the sea continuously for one year, but not for longer. It ought also perhaps to tell us the depth of water from which the mass was torn; but nothing is known as to the depth to which *Pholas* may extend —it has the reputation of occurring between tide-marks or just below, but it may extend downwards wherever there is a submarine cliff.

Though we are still unable to locate exactly these submarine ledges or fix their depth below the sea, the blocks of moorlog are so widely distributed around the Dogger Bank, and have been dredged in such large masses, that it seems clear that a "submerged forest" forms part of the core of the bank. As nothing else approaching to a solid stratum appears to be dredged over this shoal, we may assume that the moorlog forms a sort of cap or cornice at a depth of about 10 fathoms, overlying loose sandy strata, and perhaps boulder clay, which extend downward to another 10 fathoms, or 120 feet altogether. Unfortunately we cannot say from what deposit the large bones of extinct animals were washed ; they may come from the sands below the moorlog, but it is quite as probable that the Pleistocene deposits formed islands in the ancient fen—as they do now in East Anglia, Holderness, and Holland.

More than one submerged forest may be present on the Dogger Bank. The masses of moorlog are usually dredged on the slopes at a depth of 22 or 23 fathoms ; but at the south-west end it occurs on the top as well as on the slope, the sea-bottom on which the moorlog is found consisting of fine grey sand, probably an estuarine silt connected with the submerged forest, for the North Sea sand is commonly coarse and gritty.

With regard to the moorlog itself and its contents, it is possible that some of the mammals in the list, such as the reindeer, beaver, and walrus, may belong to this upper deposit; but we have no means of distinguishing them, as the bones were all found loose and free from the matrix. The insects and plants were all obtained from slabs of this peat.

The dredged cakes of peat handed to us for examination came from different parts of the Bank; but they were all very similar in character, and showed only the slight differences found in different parts of the same fen. The bed is essentially a fen-deposit of purely organic origin, with little trace of inorganic

mud. It is fissile and very hard when dry, and in it
are scattered a certain number of fairly well-preserved
seeds, principally belonging to the bog-bean. Other
recognisable plant-remains are not abundant. They
consist of rare willow-leaves, fragments of birch-wood
and bark, pieces of the scalariform tissue and sporangia
of a fern, and moss, and, curiously enough, of groups
of stamens of willow-herb with well-preserved pollen-
grains, though the whole of the rest of the plant to
which they belonged had decayed.

The material is exceptionally tough, and is very
difficult to disintegrate. In order to remove the
structureless humus which composed the greater part
of the peat, we found it necessary to break it into
thin flakes and boil it in a strong soda solution for
three or four days. Afterwards the material was
passed through a sieve, the fine flocculent parts being
washed away by a stream of water, the undecomposed
plant remains being left behind in a state for examination. These remains were mixed with a large amount
of shreds of cuticle, etc., but recognisable leaves were
not found in the washed material.

The general result of our examination is to suggest
that the deposit comes from the middle of some vast
fen, so far from rising land that all terrigenous material
has been strained out of the peaty water. The vegetation, as far as we have yet seen, consists exclusively
of swamp species, with no admixture of hard-seeded

edible fruits, usually so widely distributed by birds, and no wind-borne composites. The sea was probably some distance away, as there is little sign of brackish-water plants, or even of plants which usually occur within reach of an occasional tide; one piece however yielded seeds of *Ruppia*. The climate to which the plants point may be described as northern. The white-birch, sallow and hazel were the only trees; the alder is absent. All the plants have a high northern range, and one, the dwarf Arctic-birch, is never found at sea-level in latitudes as far south as the Dogger Bank (except very rarely in the Baltic provinces of Germany).

The plants already found are :—

Ranunculus Lingua	Betula alba
Castalia alba	,, nana
Cochlearia sp.	Corylus Avellana
Lychnis Flos-cuculi	Salix repens
Arenaria trinervia	,, aurita
Spiraea Ulmaria	Sparganium simplex
Rubus fruticosus	Alisma Plantago
Epilobium sp.	Potamogeton natans
Galium sp.	Ruppia rostellata
Valeriana officinalis	Scirpus sp.
Menyanthes trifoliata	Carex sp.
Lycopus europaeus	Phragmites communis
Atriplex patula	

Among the nine species of beetle determined by Mr G. C. Champion it is noticeable that two belong to sandy places. This suggests that the fen may have had its seaward edge protected by a belt of sand-dunes, just as the coast of Holland is at the present day.

This submerged forest in the middle of the North Sea has been described fully, for it raises a host of interesting questions, that require much more research before we can answer them. A sunken land-surface 60 feet and more below the sea at high-tide corresponds very closely with the lowest of the submerged forests met with in our dock-excavations. But if another bed of peat occurs at a depth of 130 or 140 feet at the Dogger Bank, this would be far below the level of any recently sunk land-surface yet recognised in Britain. Also, if the slabs of very modern-looking peat, containing only plants and insects still living in Britain, come from such a depth, out of what older deposit can the Pleistocene mammals, such as elephant, rhinoceros, and hyaena, have been washed?

These questions cannot be answered conclusively without scientific dredging, to fix the exact positions and depths of the outcrops of moorlog. When we remember also that beneath a submerged forest at about the depth of the Dogger Bank there was found at Tilbury, in the Thames Valley, a human skeleton; and that both human remains and stone implements have been discovered in similar deposits elsewhere, we can point to the Dogger Bank as an excellent field for scientific exploration.

The Dogger Bank once formed the northern edge

of a great alluvial plain, occupying what is now the
southern half of the North Sea, and stretching across
to Holland and Denmark. If we go beyond the
Dogger Bank and seek for answers to these questions
on the further shore, we find moorlog washed up
abundantly on the coasts of both Holland and Denmark,
and it has evidently been torn off submerged ledges
like those of the Bank. Numerous borings in Holland
give us still further information, for they show that
beneath the wide alluvial plain, which lies close to the
level of the sea, there exists a considerable thickness
of modern strata. At Amsterdam, for instance, two
beds of peat are met with well below the sea-level, the
upper occurring at about the level of low-tide, the
lower at a depth of about 50 or 60 feet below mean-
tide. That is to say, the lowest submerged land-
surface is found in Holland at just about the same
depth as it occurs in England, and probably on the
Dogger Bank also.

Below this submerged land-surface at Amsterdam
are found marine clays and sands, which seem to
show that the lowest "continental deposit," as it is
called by Dutch geologists, spread seaward over the
silted-up bed of the North Sea; but no buried land-
surfaces have yet been found below the 60-foot level
anywhere in Holland.

This appearance of two distinct and thick peat-
beds, underlain, separated, and overlaid by marine

deposits, seems to characterise great part of the
Dutch plain. It points to a long period of subsidence, broken by two intervals of stationary sea-
level, when peat-mosses flourished and spread far and
wide over the flat, interspersed with shallow lakes,
like the Norfolk broads.

The enclosed and almost tideless Baltic apparently
tells the same story, for at Rostock at its southern
end, a submerged peat-bed has been met with at a
depth of 46 feet.

On passing northward into Scandinavia we enter
an area in which, as in Scotland, recent changes in
sea-level have been complicated by tilting, so that
ancient beach-lines no longer correspond in elevation
at different places. The deformation has been so
great that it is impossible to trace the submerged
forests; they may be represented in the north by
the raised beaches, which in Norway and Sweden, as
in Scotland and the north of Ireland, seem to belong
to a far more recent period than the raised beaches
of the south of England. It seems useless to attempt
to continue our researches on submerged forests further in this direction, especially as during the latest
stages, when we know England was sinking, Gothland
appears to have been slowly rising. Those who wish
to learn about the changes that took place in the
south of Sweden should refer to the recent mono-
graph by Dr Munthe.

CHAPTER V

THE IRISH SEA AND THE BRISTOL CHANNEL

ON the west coast of Scotland, as on the east, the succession of events seems to have been quite different from that which can be proved further south. It looks as though we must seek for equivalents of our submerged forests in certain very modern looking raised beaches and estuarine deposits, such as those of the Clyde. Even when we move southward to the Isle of Man deeply submerged post-glacial land-surfaces appear to be unknown, though there is evidence of a slight sinking, and roots of trees are found a few feet below the sea-level. In the Isle of Man we still come across the modern-looking raised beaches so prevalent in Scotland though unknown in England.

The Lancashire and Cheshire coasts, with their numerous deep estuaries and extensive flats, are noted, however, for their submerged forests, sometimes seen on the foreshore between tide-marks, sometimes laid open in the extensive dock or harbour works. The Heysham Harbour excavations, for instance, were carried far below sea-level and a thin peat-bed was met with in a boring at 52 feet below Ordnance datum. Mellard Reade considered this peat once to have been continuous with an ancient land-surface seen between tide-marks. A boring is not altogether satisfactory evidence for the occurrence of a land-surface at such a depth; but if it is trustworthy it points to a subsidence of about 60 feet, an amount identical with that observed in the Thames Valley.

The estuaries of the Ribble, Mersey, and Dee tell a similar story, for on their shores and under their marshes are found some of the most extensive submerged land-surfaces now traceable in Britain. Many accounts of these have been published; but the alternations of marine with freshwater strata and with land-surfaces are so like those already described that a short account will suffice.

Carefully plotted engineer's sections will be found in Mellard Reade's papers, and his account of the succession is so interesting that it is worth quoting. He postulates two periods of elevation, alternating with three periods of depression; but in this area, as in the Thames Valley, it appears as though all the phenomena can be accounted for by one long period of intermittent depression. His generalised section of the deposits in these estuaries is as follows:—

3rd period of depression
{ Blown sand
Recent silts with beds of peat; *Scrobicularia*, occasional freshwater shells, red-deer, horse, *Bos primigenius*, *Bos longifrons*, and human skull

2nd period of elevation } Superior peat- and forest-bed

2nd period of depression	{	Formby and Leasowe marine beds; human skeleton, bones of horse and red-deer, *Scrobicularia, Tellina baltica, Turritella communis,* etc.
1st period of elevation	}	Inferior peat- and forest-bed
1st period of depression	{	Washed drift-sand (apparently no contemporaneous fossils) / Boulder clay

It may be an accidental coincidence; but it is noteworthy that both the Mersey and Thames show two main peat-beds separated by marine strata.

The forest exposed on the foreshore at Leasowe (frontispiece) is a particularly good example of these old land-surfaces, and it is often visible. It evidently once formed a wet, peaty flat on which grew swamp plants, brushwood, and some large trees. Parts of it show a perfect network of the rhizomes of *Osmunda.* This "superior peat and forest-bed" was forming when the sea was only a few feet below its present level. The "inferior peat and forest-bed" probably indicates a drier soil; but it is difficult to get at and requires fuller investigation.

The excavation for an extension of the Barry Docks, in Glamorganshire, exposed in 1895 an interesting succession of deposits, and fortunately a particularly competent observer, Dr Strahan, was on the spot to note them and their exact levels. He also obtained masses of material from each of the beds, and from an examination of the contents

of these I was able to gather a clear idea of the
changes of sea-level which had affected this part
of South Wales. The following sequence was met
with:—

1. Blown sand.
2. *Scrobicularia*-clay   Recent subaerial
3. Sand and gravel with rolled   and
shells (*Scrobicularia, Tellina, Car-*   tidal deposits.
dium, Patella, Littorina).

Strong line of erosion.

4. Blue silt with many sedges, and at the bottom
a few foraminifera.

5. *The Upper Peat Bed*, about four feet below
Ordnance datum and fairly constant in level. It
ranges from one to two feet in thickness, and where
fully developed it presents the following details:—

5 *a*. Laminated peat with logs of willow, fir
and oak, passing down into

5 *b*. Light-coloured flexible marl composed of
ostracoda with much vegetable matter.

5 *c*. Shell-marl composed principally of *Limnaea, Bythinia*, etc., with ostracoda and much vegetable matter. This seam must have been formed in
a nearly freshwater tidal marsh; it yielded *Najas
marina*, a plant now confined to Norfolk.

5 *d*. Peat with logs of oak, etc. A Neolithic
worked flint was found by Mr Storrie in this seam,
three inches below the shell-marl. This implement

is a fragment of a polished flint celt, which seems to have been used subsequently as a strike-a-light. Two bone needles are said to have been found in this peat-bed during the construction of the first Barry Dock.

6. Blue silty clay with many sedges. From five to seven feet in thickness.

7. *The Second Peat* is an impersistent brown band, a few inches in thickness, composed mainly of *Scirpus maritimus*. It suggests merely that for the time plant-remains were accumulating more rapidly than mud.

8. Blue silty clay, like Nos. 6 and 4. In its upper part, immediately under the peat bed No. 7, it contains land and salt-marsh shells, *Helix arbustorum, Pupa, Melampus myosotis, Hydrobia ventrosa.* Upright stems of a sedge, probably *Scirpus maritimus,* occur throughout this bed as through all the other silts.

9. *The Third Peat* occurs at or close to the bottom of the dock, at 20 feet below Ordnance datum. It rarely exceeds eight inches in thickness, but is persistent. In several places it is made up almost entirely of large timber, both trunks and stools of trees, while in one section roots and rootlets extended downward from the peat into a soil composed of disintegrated Keuper Marl. Mr Storrie identified oak and roots of a conifer. On washing a sample collected at a few yards' distance, I found it to consist of a tough mass of vegetable matter, principally

sallow and reed, both roots and stems. It also contained seeds of *Valeriana officinalis* and *Carex*, and elytra of beetles. There was no evidence of salt water.

At this point it will be observed that the floor of Keuper Marl rises, and Bed 9 abuts against it. Beds 10, 11 and 12 lay below the dock bottom, and were exposed only in the excavation for the foundations of walls, etc. Fortunately, Dr Strahan was able to examine a good exposure of the important part of them.

10. The section commenced at the dock bottom—that is, at the peat last described (No. 9); in the upper part it was timbered up, but at a depth of about nine feet, blue silty clay of the usual character could be seen and dug out through the timbers. This was followed by two feet of greenish sandy silt full of reeds, and containing leaves of willow, and land and freshwater shells, such as *Limnaea auricularia, Planorbis albus, P. nautileus, Hydrobia ventrosa, Valvata piscinalis, V. cristata.* The plants were *Salix caprea* and *Phragmites.*

11. Peat with much broken oak-wood, mixed with seeds and freshwater shells. The plants obtained were oak, hazel, cornel, hawthorn, bur-reed and sallow.

12. Reddish clayey gravel with land shells and penetrated by roots, passing down into red and green grits, limestone and marls. This gravel is

undoubtedly an old land-surface, lying at a depth of 35 feet below Ordnance datum. This old soil contains :—

Carychium minimum	Pupa
Helix arbustorum	Valvata piscinalis
,, rotundata	Cardium edule (two fragments—
,, hispida	probably brought by gulls)
Hyalinia	Crataegus Oxyacantha (seed)
Succinea	Cornus sanguinea (seed)
Limnaea truncatula	Quercus Robur (wood)

The examination of these deposits made it perfectly clear that the lowest land-surface represents a true forest-growth, such as could only live at an elevation clear of the highest tides; one tide of brackish water in the year would have sufficed to alter markedly the character of the fauna and flora of the deposit. Dr Strahan, assuming that the range of the tides was the same as at the present day, and noting the present highest level to which the salt-marshes reach, comes to the conclusion that 55 feet at least is the amount of the subsidence. I should be inclined to add a few feet more, in order to keep the oak-roots well clear of the highest tide during a westerly gale. An exceptional gale occurring only once during the lifetime of an oak might bank up the sea water sufficiently to kill the tree, if it grew at a lower elevation.

It may be argued that when the land stood at the higher level the range of the tides was less, and that consequently the amount of the proved subsidence may not be so great as 55 feet. The old land-surface on which the oaks grew lies, however, 35 feet below *mean tide*, so that any supposed lesser tidal range in ancient times could not make any great difference in the amount of subsidence here proved—it cannot be less than 45 feet. When, however, we notice the rapid increase in the range of the tides at the present day as the channel narrows towards Chepstow, and think what would be the probable effect of raising the whole country 50 or 60 feet, we are compelled to think that any narrowing and shoaling of the channel would have the effect of increasing, not decreasing, the tidal range at Barry Docks. In short it looks as if when the lowest submerged forest grew, the abnormal tides of Bristol may have extended further west, to near Cardiff.

Whatever may have been the exact range of the tides in these early days, it seems that the Bristol Channel points to a subsidence in post-glacial times of about 60 feet—or just the same amount as the Thames, Humber, and Mersey. The amount may have been more; but the Barry Dock sections show that it cannot have been less; we will return to the question of its maximum extent later on.

Before leaving this locality it may be well to enquire what further light it sheds on the movement of submergence, and on its continuous or intermittent character. The succession of the strata above the lowest land-surface, and the nature of their enclosed fossils, suggest long-continued but intermittent subsidence; I can see, however, no indication of a reversal of the process. The land-surface is carried beneath the water, the estuary then silts up, becomes fresh water, marsh-plants grow, and even trees may flourish on this marsh before it subsides again. But there is no sign that the strata were ever raised above the level to which ordinary floods could build up an alluvial flat. The land-surfaces seem always to have been swampy, and bed succeeds bed in fairly regular sequence, without the deep channelling we might expect to find when an alluvial flat was raised to a noticeable extent above the level of high water.

The width of the Bristol Channel makes it clear that this gulf must occupy a submerged valley of great antiquity. It becomes therefore of interest to enquire whether the wide valley is correspondingly deep, or whether its rocky floor is found at the same shallow depth as in the case of the other river-valleys which we have been considering. The wide valley may have been formed in either of two ways. It may have been excavated as a deep valley with its bottom many hundred feet below the present sea-level. Or it may have commenced as the shallow valley of a big river with exceptionally powerful tides, and as this river swung from side to side it greatly widened its valley without making it any deeper.

Possibly a deep channel may exist towards the Atlantic; but we know that none extends as far up as Bristol. Near Bristol the Severn Tunnel was carried through Carboniferous and Triassic rock, and showed that no buried channel is found much below the present one, which here happens to be scoured by the tides to an exceptional depth. The bottom of the old channel cannot be more than 40 feet below the bottom of the present channel known as The Shoots.

It may be that the Severn was once prolonged seaward as a swift river falling in a series of rapids over hard ledges of Palaeozoic rocks; but of this there is no evidence. It also does not seem probable, for all the geological indications go to suggest that west of Bristol the Channel coincides in the main with a wide area once occupied by comparatively soft Secondary or even Tertiary rocks. However this may be, we can only trace an ancient post-glacial channel cutting to about the same depth as the channels of the other rivers, and the lowest submerged land-surface of Barry Docks corresponds quite well with an alluvial flat formed when the river ran at that level. Here again we seem to find the river cutting to an ancient base-level which was about 60 feet below the present sea.

The reader may perhaps think that this point, the limited range of the upward and downward

movements in post-glacial times, is being insisted on
with wearisome iteration.   But the insistence is
necessary when we remember how constantly both
geologists and naturalists, in order to account for
anomalies in the geographical distribution of animals
and plants, bring into play such movements. The argu-
ment is constantly used, that a certain species cannot
cross the sea : therefore if it is found in an island, that
island must once have been connected with the mainland. Nature is more full of resource than we imagine,
and does not thus neglect her children. The cumu-
lative effect of rare accidents spread over many
thousand years is also far greater than may be
thought by those who only consider what has been
noted since means of dispersal have been studied
scientifically.

An examination of the south side of the Bristol
Channel need not long delay us, except for two pieces
of evidence which should not be passed over.   In
Somerset there are wide expanses of marsh land
known as the Bridgwater and Glastonbury Levels.
These greatly resemble the Fenland, and like it are
underlain by a submerged rock-platform which has
sunk in post-glacial times. But in this case we are
able to fix a definite historical date by which all movement had ceased—it may have ceased much earlier,
but we can prove that at any rate there has been no
change of the sea-level subsequent to a certain date.

The Glastonbury Levels lie at about the height of ordinary high tides, and the channels through them would still be tidal were it not for the banks which keep out the sea. Some years ago there were discovered on the surface of these marshes a number of low mounds, which on excavation proved to be the remains of a village of lake-dwellings, approached by a boat-channel, by the side of which were the remains of a rough landing stage. The dwelling-places rested on the old salt-marsh vegetation, brushwood and soil being used to raise their floors above the level of the highest tides. It is evident that when this village was inhabited the sea-level must have been the same as now, or within a foot or two of its present height. If the sea-level was then higher, the village could not be inhabited; if it were lower the channel would not have been navigable and the landing stage would have been useless. The archaeological remains found in this village prove that it belongs to a period dating about the first century B.C. or the first century A.D.

Another locality on the south side of the Bristol Channel which we must not pass without notice is Westward Ho, in Bideford Bay. There is nothing exceptional about the submerged forest at this place, but it has been carefully studied and collected from by Mr Inkermann Rogers, and it may be taken as a typical example of such deposits in the south of England.

The peaty deposits and old land-surface here seen between tide-marks are rapidly being destroyed by the sea and are now much thinner than they were a few years since. The soil on which the trees, here mainly oaks, are rooted consists of a blue clay full of small pebbles and fragments of the Culm Measure grit. Among these stones are numerous flint-flakes made by man ; but metal implements and pottery, so common in the later deposits at Glastonbury, have not been found. This ancient land-surface lies several feet below high water; it shows therefore that the latest movement of depression dates from a period between this Neolithic deposit and the Celtic lake-dwelling of Glastonbury.

The possibility of fixing an approximate date for this submerged forest, through its numerous flint-flakes and the accompanying bones of domesticated animals, makes its contents of great interest, for it shows how recently the movement has ceased—probably not more than 3500 years ago. It will be worth while therefore to give a fuller account of the contents of this soil and its overlying peat-bed.

As regards articles of human workmanship, I have seen nothing but waste flakes of flint and perhaps flint knives ; and though good implements may at any time be discovered, neither chipped nor polished tools seem yet to have been found. Human remains are represented by a clavicle.

The accompanying mammals are the stag, Celtic shorthorn, horse, dog (a very slender breed), sheep, goat, and pig, all of which, except the stag, seem to be domestic animals. Dr Chas. Andrews remarks that the ox seems to be certainly the Celtic shorthorn (*Bos longifrons*), while the small sheep is a characteristic Romano-British form, which has been described from many places, where it has been found with Roman and earlier remains.

A number of seeds were obtained from the peat which rests on this old land-surface, and it is noticeable that several of them belong to brackish water or sea-coast plants. No cultivated species have yet been found, either here or elsewhere, in even the newest of the submerged forests. The list of plants is still a small one; but it may be worth giving, to show what species can be identified. It must not be forgotten that in such deposits plants which do not possess either deciduous leaves or hard seeds leave no recognisable traces, though they may have been quite as abundant as the hazel, of which everyone notices the nuts. The seeds belong to:—

Ranunculus Flammula	Rubus fruticosus
,, repens	Callitriche
,, sceleratus	Cornus sanguinea
Viola	Sambucus nigra
Malachium aquaticum	Aster Tripolium
Stellaria media	Solanum Dulcamara
Lychnis Flos-cuculi	Ajuga reptans

Sueda maritima	Quercus robur
Atriplex patula	Alisma Plantago
Rumex	Ruppia maritima
Urtica dioica	Eleocharis palustris
Alnus glutinosa	Scirpus Tabernaemontani
Corylus Avellana	Carex 3 sp.

In this list, as is usually the case with the newest submerged forest, we find only plants that are still living in the immediate neighbourhood. Also, only such plants as are widely distributed are here found as fossils, the characteristic west-country flora being unrepresented. The reason of this limitation will be discussed later.

For various reasons, which will be explained later, it will be well before describing the submerged land-surfaces of Cornwall and the Atlantic coast, to complete the account of those surrounding our enclosed seas. We will therefore take next those bordering on the English Channel.

CHAPTER VI

THE ENGLISH CHANNEL

THE English Channel, like our other enclosed seas, is bordered on either side by a fringe of ancient alluvia and submerged forests, which however are fast disappearing through the attacks of the waves. The destruction is so rapid, and in many parts has been so complete, that we are apt to forget how altered is the appearance of the English coast. Even so recently as the time of Caesar's invasion flat muddy shores or low gravelly plains occupied many parts of the coast where we now see cliffs and rocky ledges.

We will not labour this point, which must be obvious to anyone who has noticed how little the low terrace which still fringes great part of the Sussex coast can resist the waves, and how quickly it is eaten away during storms. Any restoration of our coast-line for the time of Caesar must take these changes into account.

The material thus being removed by the attacks of the sea is partly Pleistocene gravel, partly alluvium of later date ; and the alluvial strata with their accompanying buried land-surfaces resemble so closely those already described that we need not linger long over their description.

If we commence at the Strait of Dover we are immediately confronted with clear evidence of the change of sea-level. Submerged forests are well seen between tide-marks in Pegwell Bay, and valleys with their seaward ends submerged and forming harbours are conspicuous in Kent. Owing to local conditions, the valleys are mostly narrow and steep, and the small harbours therefore soon filled up, or were lost through the cutting back of the cliffs on either side.

Possibly in Caesar's day good natural harbours were still in existence here.

Unfortunately on this part of the coast the study of coastal changes has been involved in a good deal of needless obscurity. Many writers, even geologists, make no clear distinction between loss by submergence and loss by marine erosion. We are told, for instance, that the Goodwin Sands were land about 900 years ago, and that this land disappeared during an exceptional storm. We are sometimes even told that here and elsewhere walls are still visible beneath the sea. Popular writers, to add to the confusion, have some hazy notion that these changes are connected with the existence of submerged forests or "Noah's Woods," and that these again are evidence of a universal deluge. The whole of the arguments are strangely tangled, and we must try and make things a little clearer before passing on. An understanding of the changes which have taken place on this part of the coast is needed for historical purposes, and still more needed if we make a study of the origin of the existing fauna and flora of Britain.

One of the crucial questions, both for the naturalist and archaeologist, is the date at which Britain was finally severed from the Continent. Did this happen within the range of written history, or tradition? Or if earlier, did it take place after or before climatic conditions had become such as we now experience? For the proper understanding of many different problems it is essential to settle this point.

It is scarcely satisfactory to read history backwards, though geologists are often compelled thus to work from the known to the unknown. We will therefore not in this case ask our readers to follow us through the detailed evidence and arguments which have enabled geologists stage by stage to reconstruct the physical geography of this part of Britain as it was in days before written history. They must take this preliminary work for granted, and allow the description of the changes to be taken in their correct historical order.

We need not go back far geologically. In late Tertiary (probably Newer Pliocene) times there was a ridge of chalk joining the range of the North Downs to the corresponding hills of France; but the divide between the North Sea and the English Channel was low at this point. Afterwards, during the Glacial Epoch, when an ice-sheet accumulated and blocked the northern outlet of the North Sea, the water was ponded back in the southern part. There was no easy outlet northward for the water of the Rhine and other great rivers, so the level of the North Sea rose slightly till it overflowed this low col and cut an outlet where lies the present Strait of Dover.

The general sea-level during this period of

glaciation seems to have been a few feet higher than that of the present day, for glacially transported erratics are found strewn over the flat coastal plain of Sussex. One erratic block, probably derived from the Channel Islands, was discovered under the loess as far east as Sangatte cliff, close to Calais. The icy English Channel must therefore have met the icy North Sea some time during the Glacial Epoch.

Some time after the cold had passed away there came in the period with which this book deals—when the lowest submerged forest flourished, on land now 50 or 60 feet below the sea. This elevation of the land, as already shown, converted a great part of the North Sea into a wide alluvial plain. At the same time it raised above the sea-level and obliterated the newly-formed strait, leaving it in all probability as a shallow valley sloping both ways and filled up with alluvium. The Strait of Dover was again a watershed, or perhaps its position was occupied by a small stream, which may have flowed in either direction.

Thus the work done during the Glacial Epoch was almost cancelled and had to be done again; but now there was merely a low narrow divide of chalk and a strip of marsh between the two basins, and the chalk ridge was steadily being attacked by the waves of the sea from the west.

When subsidence again set in the strip of alluvium was soon submerged and the two seas again met; but

in all probability for a long time the Strait was only
a narrow one, over which animals could easily swim
Then tidal scour, deeper submergence, and the action
of the waves did the rest, so that ever since that time
the Strait of Dover has been getting steadily wider
and wider, and also deeper. Its bottom is to a large
extent composed of bare chalk with patches of gravel;
and the movement of this gravel during storms, combined with the action of boring molluscs must slowly
eat away the chalk far below ordinary wave-action.

The above explanation is needed, for it will not
do to take existing soundings, and say that all the
sea-bottom below a certain level, corresponding with
a particular submerged forest, was then sea and all
above was then land. This is an easy way of recon-
structing the physical geography ; but it may be a
very misleading one. A little consideration will show
that whilst in large areas sandbanks have accumulated
to a great thickness, in other areas, of which we know
the Strait of Dover is one and the Dogger Bank a
second, there has been much submarine erosion,
which is still going on. In neither case is it safe
entirely to reconstruct the ancient contours from the
present-day soundings.

Even such a gigantic feature as the continental
platform, which ceases suddenly at a depth of
100 fathoms, is in all probability in the main a
feature formed by the deposition of sediment during

long ages. Its outer edge marks, not the limit of
some ancient continent, but the limiting depth at
which gentle wave-action has been felt, and beyond
which the sediment cannot be carried.

After this necessary digression we must return to
our study of the actual evidence for such changes of
sea-level in the English Channel. It has been pointed
out already that for this purpose the present depth
below sea-level of the rocky floor of the Strait cannot
in itself be accepted as sufficient evidence. Nor can
the depth at which rock was met with under the
Goodwin Sands ; though here a cylinder was sunk
75 feet before it reached the chalk. Unfortunately
no record of the strata passed through seems to have
been preserved, though it is perhaps implied that
nothing but sea-sand was penetrated.

Romney Marsh is a wide alluvial flat occupying
a silted-up bay, the floor of which in places lies at
least 70 feet below sea-level. There are here unfortunately no extensive excavations for docks, and
all we can say is that the few borings which have
penetrated the alluvial strata prove the existence of
a slightly undulating rock-surface below. In short
Romney Marsh appears to be a submerged flat-
bottomed open valley, like that which we have
already seen underlies the marsh deposits of the
Fenland.

In the case of Romney Marsh, however, it is

doubtful whether submerged land-surfaces would be found at any great distance from the rising ground. There is a striking peculiarity about this marsh; it only lies partly in a bay, the greater part of the area consisting of alluvial flats which have accumulated during recent centuries behind the projecting shingle beaches of Dunge Ness. In short, the marsh steadily gains on the sea, is advancing into fairly deep water, and the parts near the Ness may be underlain by marine strata right down to the Wealden rocks below. The rock floor was met with at 58 feet below the marsh at Holmston Range, not far from the Ness; but we have no information as to the character of the strata passed through before this floor was reached. In all probability this floor at 58 feet would be proved to be part of a true land-surface, could we examine it.

Near Hastings the submerged forests have long been known, and are often exposed on the foreshore between tide-marks. They contain antlers of deer, leaves, hazel nuts, acorns, and oak wood.

Then we come to Pevensey Level, which is another of the submerged and silted up wide flat-bottomed valleys, such as we have so often met with. But as we have no details as to strata underlying this marsh we must pass on.

Along the Sussex coast west of Beachy Head a series of south-flowing rivers reaches the sea, each cutting through the high chalk-hills of the South Downs. We need not discuss the origin of these peculiar courses, which date back to the period when the central axis of the Weald was uplifted; that discussion would take too much time, and is here unnecessary. We are now only concerned with the later stages of the evolution of these river-valleys, each of which yields striking confirmation of the view that a sinking of the land has taken place in comparatively modern times.

At the present day the tidal part of each of these rivers extends right through the Downs into the lower Wealden area, and it is obvious that their valleys tend to silt up, not to deepen, and scarcely anywhere to become wider. When we examine further we find that the true valley-bottom lies far below the present alluvial flat; though the scarcity of borings and the uncertainty of many of the records make it difficult to say exactly how deep it lies.

If we follow these rivers upwards we find that in each case the alluvial flat widens out greatly after we have passed the chalk-hills and reached the clay lands beyond. These wide flats, according to old ideas, were formed by the swinging from side to side of the stream, which thus gradually widened its valley in the softer strata. If this were the case in these instances, we should find a solid floor beneath each marsh at a depth not exceeding that of the present river-channels. The rivers, however, are not now cutting into rocky banks or flowing over beds of Secondary strata; they are flowing sluggishly in the middle of alluvial flats, which tend to silt up with every flood or exceptionally high tide.

Thus all the evidence seems to show that marshes like those near Lewes and Amberley Wild Brook have originated through the submergence of flat-bottomed valleys cut in soft strata. The ponding back of the muddy tidal water would soon lead to the silting-up of any shallow lakes left after this submergence.

When the land stood 70 or 80 feet higher than it does now, the country must have looked very different. The rivers then traversed the chalk downs through V-shaped comparatively narrow valleys; but these valleys opened out in their upper reaches, where they crossed the Gault and Weald Clay. If we could lay bare the true floor of the valley, we should see however that there is always a steady and fairly regular fall seaward, just as there is in the part of its course which lies above the influence of the submergence, which is felt for some 10 or 12 miles from the sea. Except on this theory of recent submergence it seems impossible to account for these curious marshes, with tributary valleys obviously plunging sharply beneath them on either side; they are quite unlike the undulating flats which occur higher up.

The flat of Selsey Bill yields evidence of sub-merged land-surfaces opposite each of the shallow valleys; but here we meet with the same difficulty which confronted us in the Thames Valley and on the east coast. Pleistocene land-surfaces and alluvial deposits of early date are seen on the foreshore side by side with the more modern Neolithic alluvium and submerged forests. Unless great care is taken it may be thought that the well-preserved bones of rhinoceros and elephant, and the shells of *Corbicula fluminalis*, come from the same alluvium that yields Neolithic flint-flakes, or that plants such as the South European *Cotoneaster Pyracantha* flourished in Britain up to this recent date. Except for the sake of warning against these sources of error the submerged forests of Selsey Bill need not detain us.

Still travelling westward we next arrive at the series of tidal harbours opening into Spithead, Southampton Water and the Solent. All of these are obviously continuations of the valleys which lengthen them inland; and this is amply proved by dock excavations and borings.

Even Southampton Water and the Solent them-selves are nothing but submerged valleys. A well at the Horse Sand Fort—one of the iron forts which rises out of the sea at Spithead—showed a band of compressed vegetable matter, probably an old land-surface, more than 50 feet below high-water level, the floor of Eocene strata not being met with till 98 feet below high water was reached. In this case, however, the strata below 50 feet seem, from the published description, to be of marine origin.

The well at Norman Fort is stated to have penetrated to a depth of 127 feet below the sea before Eocene strata were reached; but in this case the lower strata were of marine origin, and the only land animal recorded was a jaw of red deer, found apparently between 80 and 90 feet down. These deep channels may be relics of the very ancient (Tertiary) Solent River, and were probably arms of the sea till they were sufficiently silted up for the lowest submerged forest to grow.

We have not yet sufficient data, nor is it necessary to our purpose, to give a detailed reconstruction of this interesting area during the successive stages of elevation and depression. During the time when the lowest of the submerged forests flourished the Isle of Wight was connected with the mainland where the Solent now narrows about Yarmouth, and probably for some distance westward. This connexion was kept up till comparatively recent times, only breaking down finally a short time before Caesar's invasion.

In early Neolithic times the ancient Solent Valley

had already been decapitated by the inroads of the
sea west of the Needles, and the remains of this big
valley were occupied by a small river flowing eastward through the middle of the present Solent. In
its course it received numerous tributaries on either
side. It probably opened out into an estuary where
it joined Southampton Water, and so continued to and
beyond Spithead, receiving other tributaries from
the valleys now occupied by Portsmouth Harbour,
Langston Harbour and Chichester Harbour.

In time we may be able to make a more complete
reconstruction of the physical geography of this area
for definite dates; but the point now to be insisted
on is that the Isle of Wight was part of the mainland
up to quite recent times, so that its fauna and flora
could readily pass backwards and forwards without
crossing the sea.

Perhaps to the geographer or geologist one of the
most striking confirmations of a recent submergence
affecting this part of England will be found in the
strange series of enclosed harbours extending from
Chichester westward to Fareham. These harbours
are not each distinct and separate; all of them have
cross connexions in the form of shallower channels
some four or five miles inland from Spithead. I have
often been asked what is the meaning and origin of
these peculiar harbours, which are not forming or
widening now, but rather tend to silt up.

The origin of these harbours is quite easy to understand, if we admit the recent sinking of the land, and for this we will presently give abundant evidence. On any other hypothesis these inosculating water-ways must seem hopelessly confused and inexplicable. Sea and waves do not erode enclosed basins such as these.

Granting the submergence, we see that each of these harbours must once have been a shallow valley; but this does not account for their basin-like shape and their cross connexions. For the reason of these peculiar features we must look at the map by the Geological Survey showing the superficial deposits. It will then be seen that all this part of Hampshire shows a widespread sheet of gravel and gravelly loam which slopes gently seaward and passes below the sea at Spithead. Northward the gravel rises, and the soft Eocene and Cretaceous strata appear beneath the gravel between tide-marks at various places toward the northern ends of these harbours.

The waves of the sea can remove loose gravel as readily as clay, and we see that on this coast wave-action is practically confined to the low cliff facing the sea and does not affect the interior of the harbour. But it is well known to geologists that a sheet of coarse angular gravel such as this, notwithstanding its looseness, is much less readily attacked by a small stream than is a surface of hard clay or

even chalk. Thus plains of Tertiary deposits capped
by gravel, under the action of rain or rivers develop
into gravel-capped plateaus and hills, which fall
abruptly into open flat-bottomed valleys. The denudation takes place at the edge, where the gravel rests
on the Tertiary strata and numerous springs are
given out ; there is scarcely any denudation in the
gravel flat, and unless the height of the land is
considerable there is no great amount of denudation
in the flat bottom of the valley.

Thus there is a tendency for the valley to widen
out on every side, wherever the gravel rests on
impervious or soft strata. But where the gravel
plunges below the water-level, as it did at the
entrance to each of these harbours, the valley
narrowed, for there were no landslips, the drainage
was subterranean, and the stream could not readily
remove the large flints.

The widening of the valleys, where they were cut
in soft strata, led to the development of small lateral
valleys to the right and left, leaving only narrow
divides between their head waters and those of the
next valley. When the land sank these divides were
flooded, and so were developed the shallow cross
connexions, much as we now see them.

In order that it may not be imagined that this
reconstruction is merely hypothetical, it will be as
well to give some evidence that such an elevation

and submergence did take place in this district as in others. We cannot in this little book deal with the whole of the evidence, so we will take the Southampton Dock excavations as sufficient to prove our point, condensing the account from the *Geology of Southampton*, published by the Geological Survey.

The general section at Southampton Docks is as follows, though the thickness varies considerably at different points, and the greatest depth of the old valley has not yet been proved :—

	Feet
Estuarine silt	20
Peat, old vegetable soil, or tufaceous marl; ox, pig, horse, pine, beech, birch, oak, and hazel	variable up to 17
Gravel, with reindeer	10 or more

The bottom gravel is apparently of Pleistocene date, though it may include also a basement bed belonging to the newer deposits. T. W. Shore recorded from the peat above the gravel a fine stone hammer-head of Neolithic date and worked articles of bone, but no instruments of metal were found. The associated marl was full of freshwater shells.

Poole Harbour tells a similar story, and evidence of this submergence is seen in the various submerged forests found along the Dorset and Devon coasts, opposite the mouths of the valleys. These rocky coasts are, however, so different from those we have just been describing, that they will more conveniently be treated of in a separate chapter.

CHAPTER VII

CORNWALL AND THE ATLANTIC COAST

ON travelling westward into Cornwall we enter a region which is extremely critical in any enquiry as to the amount of change that the sea-level has undergone. As long as we were dealing with ancient river-channels opening into enclosed seas, like the North Sea or Irish Sea, it might be said that the depth to which the channel was cut was not necessarily governed by the sea-level. It might be governed by the level of an alluvial plain, which then extended for hundreds of miles further, and had its upper edge high above the sea-level.

This cannot be said in Cornwall, for there the sea-bed shelves rapidly into deep water, and the coast would not be far away, even were the land raised 200 feet or 300 feet. The rivers then as now must have flowed almost directly into the Atlantic Ocean, and their channels must then as now have cut nearly to the sea-level of the period.

The Cornish rivers yield most valuable information. It so happens that many of them bring down large quantities of tin ore from the granitic regions, and this ore being very heavy tends to find its way to the bottom of the alluvial deposits, out of which it is obtained in the same way as alluvial gold in other countries. On following this detrital tin ore downwards towards the estuaries the "tinners" or alluvial miners found in many cases that a rich layer descended lower and lower till it passed well below the sea-level in some of the ancient silted-up valleys. Some of these tin deposits were so rich that it paid even to divert the rivers, dam out the sea, and remove the alluvium to considerable depths in search of the ore.

These excavations for tin produced most interesting and continuous sections of the alluvial deposits, and if only they had been examined more thoroughly and scientifically they would have thrown much light on the questions we are here considering. Unfortunately all the deeper excavations were made in days when all ideas as to the origin of "diluvial" deposits were so tinged with theories as to the effects of a universal deluge, that many of the most interesting points escaped notice. The last of these "stream works" was closed many years ago.

Notwithstanding the early date of these excavations, some most interesting observations were recorded ; and though they make us long for fuller details and regret the loss of many of the objects referred to, we must be grateful that so much was noted, and by such careful observers. This entire removal of the old alluvial deposits—for the tin usually occurs concentrated in the bottom layers—

showed clearly that in Cornwall, as elsewhere, old land-surfaces can be found far below the sea-level. The shape of the valley-bottom, and the rapid lessening of its fall as the coast is approached, in several cases point clearly to the proximity of the sea, and show that its ancient level must have been about 70 feet below present tides.

Here it may be pointed out that as the sea-level is approached the steady seaward fall of a rocky V-shaped valley quickly lessens, changes to a gentle slope, and then to a flat, more or less wide according to the length of time during which the river has been kept at the same level, and could only swing from side to side, without deepening its bed. In Cornwall there is a definite limit below which the erosion of the valleys has not gone, and at this level the valley widens and flattens as it does elsewhere.

The eastern border of Cornwall is formed by the extensive harbour which receives the Tamar, Tavy, and Plym, and this harbour is obviously nothing but a submerged seaward continuation of the combined valleys eroded by these rivers. The rivers, it must be remembered, though short, receive great part of the drainage of Dartmoor, where the rainfall is excessive; they are therefore very liable to floods. These streams also bring down much coarse gravel and sharp granitic sand, so that their erosive power must be exceptionally great during floods. It seems therefore that the scour would always have been sufficient to keep open a channel well below low-water level.

No stream tin has been worked in the Plymouth estuary, so that we cannot point to any continuous sections in the ancient alluvial deposits, such as are found further west. These tin deposits, however, date in the main from a period somewhat earlier than that with which we are now dealing. They were probably swept down from Dartmoor when floods were far more severe, during the annual spring melting of the snow during the Glacial epoch. Unfortunately, also, the lately finished harbour works at Devonport proved the existence of only modern alluvium, without any submerged forests.

Before dealing with the rivers which flow into Plymouth Sound it is necessary, however, to say a few words about the harbour itself and its origin. Plymouth Sound and the various submerged valleys which open into it illustrate well both the continuity of geological history, and the great difficulties which await us when we deal with valley erosion which in part dates far back into Tertiary times. The Sound is not merely a submerged continuation of the Pleistocene valleys, and between this wide gulf and the narrow valleys there is a curious want of continuity. We do not know the true depth to the rocky floor; but at two places just outside the mouths of the estuaries deep hollows are scoured through

6—2

the sands. One of these, just outside the Hamoaze, or estuary of the Tamar, shows a rocky bottom at 150 feet, probably the true rock-floor of that part of the Sound. The other hollow, 132 feet, is just outside the Cattewater ; but does not reach rock.

It is obvious that these depths, both of which are measured from low water, show a depression of the rocky floor of the Sound far greater than we meet with in ordinary Pleistocene valleys ; but at present we have no means of proving the true date of this depression. It represents not improbably a Tertiary basin, like that of Bovey Tracey, which also descends several hundred feet below sea-level. In favour of this view we can point to the occurrence of a small outlier of Trias in Cawsand Bay, which certainly suggests that the Sound represents an area of depression or synclinal basin, rather than a mere submerged valley. It has also been stated that relics of Tertiary material are still to be found in the limestone quarries of Plymouth; but for this the evidence is not altogether satisfactory.

It may be asked, What practical difference does it make, whether or no the Plymouth Sound were originally a Tertiary basin, for no Tertiary gulf could now remain open ? If we were dealing with an area of soft rocks, like the Thames Valley, or with an enclosed sea, this objection would hold. Around Plymouth, however, the Palaeozoic rocks are

extremely hard, and can resist for ages the attacks of
the sea ; but loose Tertiary material, or even Triassic
strata, would readily be swept away by the heavy
Atlantic swell and by the scour of the tides, until
they were protected by the building of Plymouth
Breakwater.

There is a general impression that marine action
cannot go on much below low water ; but this is
altogether a mistake. Tidal scour may go on at any
depth, provided the current is confined to a narrow
channel, so as to obtain the requisite velocity. If in
addition there is a to-and-fro motion, such as that
caused by the Atlantic swell at depths of at least
50 fathoms, the actual current required to remove
even coarse sand need only be very gentle. The
oscillation in one direction may not reach the critical
velocity ; in the other this velocity may just be
exceeded; the movement, therefore, of the sand
grains may always be in one direction, especially
if the courses taken by the ebb and flood tides
do not coincide, or their velocities differ.

How does this apply to the origin of Plymouth
Sound ? The mere fact that opposite the mouth of
the Tamar a pit has been scoured to a depth of
150 feet, and opposite the Hamoaze another to 132 feet
below low water, and that these pits are kept open,
notwithstanding the enormous amount of sediment
brought down by these rivers, proves that tidal scour

is now going on, or was recently going on, at depths
of 25 fathoms at least in confined parts of Plymouth
Sound. Similar troughs occur at even greater depths
near the Channel Islands, where the tidal scour is
very great, and in the Bay of Biscay coarse sand is
moved at depths of at least 100 metres.

It is necessary to make this digression as to the
effects of tidal scour, for we are sometimes told that
the various basins, troughs, and channels shown on
the charts represent submerged land-valleys, and
thus prove enormous changes of sea-level in modern
times. How a submerged valley in a narrow sea with
sandy bottom, like the English Channel, could remain
long without silting up is not clear; the sandbanks on
either side should tend to wash into and fill up the
hollows. The troughs, however, all coincide with
lines of tidal scour; they do not continue the lines of
existing valleys, unless these valleys are so large as
to produce a great scour, and unless this scour is
aided by the oscillation of the waves. A glance at
the Admiralty chart will show that no submerged
channel crosses the direction of the tidal scour or of
the Atlantic swell; the channels are scoured where
tide and swell act together.

We conclude therefore that Plymouth Sound probably represents a basin once filled with soft Tertiary
and Secondary deposits, and that these soft deposits
were cleared out by the sea, leaving the rocky floor

of the basin bare at a considerable depth below sea-level. In part the basin has now silted up again ; but we may fairly consider that at the time of greatest elevation, when the submerged valleys were being eroded, the depth of water in the Sound was much the same as it is now. Then as now the rivers seem to have discharged into a wide open gulf occupied by the sea.

However this may be, we see now a series of deeply trenched valleys, partly submerged and all opening into a wide and deep bay. The e valleys do not now show rocky bottoms gradually sloping into the open harbour. The rock floor ceases several miles up and gives place first to an alluvial flat and then to an arm of the harbour. Like all the other valleys with which we have been dealing they cut to a definite base-level, approximately that of the sea, and the parts below that level are rapidly silting up.

Fortunately a large series of bridge-foundations has shown well the character of these valleys, where the rocky floor passes beneath the sea-level, and the late R. H. Worth gave an excellent series of sections across them. He took their contours to be evidence of glaciations. In this I cannot agree with him ; but think rather that the extraordinary flatness of the valley-bottoms, and especially the uniform depth to which they were excavated, point to the attainment of a definite base-level.

Commencing with the most easterly of the rivers which enter the Sound, we find that the Laira Railway Viaduct, across the Cattewater, proved a breadth of 212 feet at the centre of the channel, with the rock-floor practically level at 87 feet below low water; no V-shaped valley or gorge was met with. At Saltash the foundations of the bridge show the depth to the rock-bottom to be 75 feet; but the viaduct across the Hamoaze is about three miles higher up the river than the Laira Viaduct. The Tavy Viaduct, nearly two miles further from the sea, shows a width of 240 feet of practically level rocky floor at 67 feet below sea-level. Thus all this evidence is consistent with the existence of a series of wide open flat-bottomed valleys, now partly submerged, with a fall of about five feet in the mile. This is about the fall necessary for even a rapid river flowing through a flat so full of boulders and coarse gravel as this must have been. It must not be forgotten also that this five feet in the mile is the general fall of the valley-bottom, not of the water, and that a river winding from side to side would have about half or one-third of this fall. The slope was probably just sufficient to keep the channel clear and let the water escape.

We may take it, therefore, that the ancient valleys opening into Plymouth Harbour cut to about 100 feet below mean tide, as do the Thames and Humber, and that this was the measure of the greatest elevation of the land in Pleistocene times, for these valleys opened suddenly into a sea of considerably greater depth. A word of explanation is still required as to the meaning of the extremely flat rock-bottom, for one might have expected more of a U-shaped or V-shaped valley, unless the period of stationary sea-level were very long.

Owing to the great rush of water from Dartmoor during floods, and the enormous amount of coarse gravel swept down, the erosive power of these streams is very great. This was greatly exaggerated during the Glacial Epoch, to which the formation of the tin-ground and of the flat bottom belong. The melting of the snow in spring must have caused far more severe floods than we now see, and these floods must have brought down large quantities of river-ice heavily charged with boulders of hard and angular metamorphic rocks, such as would erode and trench in a way that does not now happen. Thus as the river changed its course or swung from side to side according to the varying amount of water, the ice-laden water must have had an erosive power more like that of a Canadian river in spring than like anything we now see in Britain. The wide and deep flat-bottomed trench need not have taken any enormous length of time to form, for river-ice and anchor-ice were constantly at work removing the loose material and laying bare the rock-face so that it could be again attacked.

The period of exceptionally rapid erosion and of low sea-level above postulated must be our starting point in Devonshire and Cornwall as elsewhere, for it fixed the shape and depth of the submerged valleys over wide areas. This erosion came somewhat earlier than the growth of the submerged forests; but it is impossible to treat of any particular period of history without some mention of what has gone before and led up to it. I may say also that I doubt whether there is any such great gap as is commonly supposed between the Glacial period and later times.

Unfortunately the succession of the newer deposits in the submerged valleys near Plymouth appears never to have been worked out, attention having been concentrated on the contour of the rocky floor. The recently completed Devonport dock excavations, which I examined, showed only very modern alluvium and silted-up channels with logs of wood cut by metal tools. Submerged forests do not appear to have been met with.

Though Plymouth Harbour has not yielded much information concerning the particular period with which we are dealing, it is important as fixing the maximum amount of elevation to which the land was subjected in Pleistocene or more recent times. We will now turn to the Cornish stream-tin works, which give more detail as to the later changes; we regret however that these most interesting excavations were closed so long ago, for various points were noted about which we should like further information, and this is not now obtainable. The old diluvian hypothesis has much to answer for in the long neglect of those modern strata which help to tie on geology to archaeology and history.

By far the best account that has come down to us of a Cornish tin stream-work carried below the sea-level, is that written by J. W. Colenso in 1829. Colenso had unusual opportunities for watching the works—apparently either as manager or owner—and he showed a most exceptional ability to note scientific points, such as were generally overlooked 90 years ago. It should be remembered that even in days before Lyell wrote we had in the Cornish tinners a class of men whose everyday occupations led them thoroughly to understand the action of running water. Their daily bread depended on their power to calculate where the ancient flood must have left the heavy tin-ore, where the barren ground would be found, or where old silted-up channels might be sought for. In their arrangements for diverting the streams in order to work the alluvial deposits, and for washing and concentrating the tin-ore, they were constantly brought face to face with the action of running water. When the buried tin-ground yielded anything abnormal the tinner recognised the effects of exceptional floods, of eddies behind boulders, or of obstructing ledges.

Where he thought he saw the action of the deluge
we may be pretty certain that he was dealing with
something truly exceptional and outside his experience of the effects of a mountain torrent. He was
not using the word as a cloak for ignorance or excuse
for indifference, as was so often the case with the
geologist of that day. Unfortunately most of the
tinners could not write.

Colenso's account is entitled *A Description of
Happy Union Tin Stream-work at Pentuan.* Pentuan
lies at the mouth of the St Austell River, a rapid
stream, much liable to sudden floods, which drains part
of the granite and metalliferous region of St Austell
Moor. The conditions are ideal for bringing down
large quantities of the decayed granite which contains
the tin-ore. This material was alternately weathered
and broken up, and so sluiced with flood-water as
to wash away the lighter quartz and felspar, thus
concentrating the tin-ore, with a small amount of
gold-dust and small gold nuggets, in the bottom
layer.

The alluvium of the St Austell River was therefore
so profitable to work that every channel was followed
upwards into the Moor, and the main valley was
followed downward towards the sea. But as the
coast was approached the rocky floor sank below the
sea-level, so that this part was left till last, for it
needed the diversion of the river and much pumping

to get rid of the water. This, scientifically, is a fortunate circumstance, for of the earlier workings in the higher part of the valley no good accounts have come down to us.

The river is only a small one and its catchment area is very limited; it has therefore a rapid fall, amounting to 30 feet in the mile between St Austell and the sea. With this fall the valley is still silting up and its alluvium rising, principally through the abnormal amount of sediment and granitic sand sent down by the china-clay works. If we take the fall of the buried channel, this amounts to about 45 feet to the mile, for the rock-floor at Pentuan lies about 60 feet below the sea-level. This rock-floor is composed of hard slates.

The successive deposits met with above the slate were as follows, commencing with the lowest:—

(*a*) *The tin ground*, or stratum in which the whole of the stream-tin is found. It lies on the solid rock and is generally from three to six feet thick, sometimes even ten feet. It extends across the valley, except where turned by a projecting hill or rock, when it is found to take the supposed ancient course of the river, which is generally under the steep bank opposite. This last observation (often made by tinners) is important, for it suggests that the heavy tin-ore was brought down by exceptional floods, such as would swing violently to the outer side of the curve, and there cut a steep bluff, under which would be left the heaviest gravel. This observation and the noteworthy absence of any contemporaneous animal remains in the tin-ground, suggest that the bottom layer may date back to Pleistocene times, when the climate was colder and floods more violent.

It is not clear how far seaward the valley may then have extended; probably not more than half a mile at most. The tin ground was worked near Pentuan for 1400 yards along the valley, and averaged about 52 yards in breadth. So here again we meet with a fairly wide flat-bottomed valley, not a narrow V-shaped gorge; we may therefore take it that the base-level had been reached and that this base-level was identical with that met with in the rivers which open into Plymouth Sound.

(b) On the tin ground were rooted numerous oaks, which had grown and fallen on the spot. Their timber was so sound that Colenso applied one of the trees to make the axle of a water-wheel, and his comment on this is excellent. "It appears to me likely that at this period, the rising of the sea had so far checked the current of the river as to prevent its discharging the mud and sand brought down with it; thus the roots were buried [submerged?] to a considerable depth, and the trees killed, before the timber underwent its natural process of decay." At one spot he records finding oysters still remaining fastened to some of the larger stones at the top of the tin ground and to the stumps of the oaks.

Then comes a stratum of dark silt, about 12 inches thick, with decomposed vegetable matter, and on this a layer of leaves of trees, hazel nuts, sticks and moss for 6 or 12 inches more. This layer of vegetable matter is about 30 feet below the level of the sea at low-water and about 48 feet at spring tides. It extends with some interruption across the valley.

The point is not made quite clear in Colenso's account, but apparently there is no marine deposit between the "tin ground" and the peat, the oyster-bed above mentioned representing the base of bed *c*, which at that point has cut through the peat, so as to lay bare part of the gravel and some of the oak-stumps rooted in it. So far, wherever we have a carefully noted section of the lowest deposits in these valleys, the tin ground or the gravels are directly succeeded by a growth of oak trees. It looks as though the climate ameliorated, the more violent floods ceased, and an oak forest grew across the alluvial flats, without there being any, or much, change of sea-level.

(*c*) Above the vegetable matter and leaves (*b*) was found a "stratum of sludge or silt" 10 feet in thickness. It showed little variation except from a brownish to a lead colour. "The whole is sprinkled with recent shells, together with wood, hazel nuts,

and sometimes the bones and horns of deer, oxen, etc.
The shells, particularly the flat ones, are frequently
found in rows or layers; they are often double or
closed, with their opening part upwards." From
Colenso's account it seems probable that this bed was
a marine silt with *Scrobicularia* and cockles in the
position of life. He goes on to say that "There has
been recently found imbedded in the silt, about two
feet from the top, a piece of oak, that had been
brought into form by the hand of man; it is about
six feet long, one inch and a half broad, and less than
half an inch thick; this is the greatest depth at which
I have ever seen any converted substance. It appears
to have floated in the sea, as at one end, which is
much decayed, a small barnacle has fixed its habita-
tion."

(*d*) *A stratum of sea-sand*, about four inches in
thickness; this is easily distinguished from the river-
sand, being much finer, and having always more or
less shells mixed with it.

(*e*) *Silt* two feet, with concretions containing
wood and bones.

(*f*) *Another stratum of sea-sand*, 20 feet in
thickness. In all parts of this sand there are timber
trees, chiefly oaks, lying in all directions; also remains of animals such as red deer, "heads of oxen of
a different description from any now known in Britain,
the horns of which all turn downwards." Human

skulls were also found near the bottom of the sand,
and one of these with other fossils was presented by
Colenso to the Royal Geological Society of Cornwall.
In the upper part of this sand nearer the mouth of
the harbour, the bones of a large whale were found.
The sea at this time seems to have extended about a
mile up the valley.

(g) *A bed of rough river-sand and gravel, here
and there mixed with sea-sand and silt.* About 20
feet in thickness. In this sand was found "the remains of a row of wooden piles, sharpened for the
purpose of driving, which appear to have been used
for forming a wooden bridge for foot passengers: they
crossed the valley, and were about six feet long; their
tops being about 24 feet from the present surface—
just on a level with the present low water at spring
tides. Had the sea-level been then as now, such a
bridge would have been nearly useless."

At Wheal Virgin, which was the upward extension
of the Happy Union works, about a mile higher up
the valley than the bridge just mentioned, the tin
ground was only 32 feet from the surface. Here
Colenso mentions seeing "on the surface of the tin-
ground two small pieces of oak, with artificial holes
in them: and there were near them several oak stakes,
sharpened and driven into the ground, and supported
by large stones. Near the same spot has been found
a substance resembling the ashes of charcoal." This

account suggests a fish or otter trap of some sort and the charcoal below the sea-level suggests that it must date back to at least as early a period as the submerged bridge. It is a great pity that antiquaries were not at that period more alive to the great interest of these finds.

Carnon stream-works, on a navigable branch of the Fal, showed a very similar section, for below about 54 feet of alternating sand and silt was found, according to Henwood, a bed one and a half feet thick of wood, moss, leaves, nuts, etc., a few oyster shells, remains of deer and other mammals, and some human skulls. Below this came the tin ground varying in thickness from a few inches to 12 feet. Here also no organic remains were found in the tin ground itself.

The above records may be accepted as giving fair samples of the deposits which now fill the lower parts of the submerged valleys of Devon and Cornwall. These valleys were all at one time long creeks or arms of the sea, navigable for a considerable distance inland and affording a fine series of sheltered harbours at short distances apart. A few of these harbours were so deep and large that they have not yet been obliterated, as is seen in the case of the Dart, the branches of Plymouth Harbour, the Fal, the Gannel, etc. A rapid silting-up is, however, now going on, greatly aided by the refuse from the mines and china-clay works. In the days whilst the subsidence was in progress Cornwall was essentially a country of fjords, though now good harbours are few or blocked with sandbanks.

The abundance of sheltered creeks must have had considerable influence on the manner of living of the inhabitants; but it is noticeable that though many acres of the silts have been removed in tinning, and a good many human remains have been found, there is no mention of boats. This absence of any record of boats in any of the marine silts associated with or below submerged forests cannot be an accident; for old boats and dug-out canoes are constantly being discovered in later alluvial and fen deposits. It looks as if in those early days man had either no boats, or only used coracles of skin and wicker, such as would entirely decay and leave no trace.

It may be remarked that the higher submerged forest, that lying just about low-water level, is not recorded in the deep excavations at Pentuan and Carnon, though these old land-surfaces are so conspicuous on the foreshore opposite every smaller creek, when the sea happens to scour away the sand and beach. A little consideration will show the reason of this difference. The extensive stream-works of Pentuan and Carnon happen to lie at the mouths of two of the larger and deeper creeks, in which silting-up could not keep pace with the subsidence.

7—2

Thus the seaward ends were continuously occupied by
sea, from the time when the oak-forest sank right on
into historic times, and over this deeply buried oak-
forest we only find alternate layers of silt and sea-sand.
Evidence of the later submerged forest, however, is
not entirely wanting, for the submerged wooden bridge
or causeway of Pentuan must belong to the period
when the trees seen on the foreshore elsewhere were
flourishing well above high-water mark.

The submerged forests seen on the foreshore in
western Cornwall are so like those exposed elsewhere
that there is no need for a full description, were it
not that they have become so connected with ancient
legends of Lost Lyonesse, a country which is supposed
to have joined the Land's End to the Isles of Scilly
somewhere about the date of King Arthur and Merlin.
To what extent these stories are due to observation
of the submerged forests and of the rapid waste of
land in Mount's Bay, supplemented by a vivid Celtic
imagination, which saw "the tops of houses through
the clear water," is doubtful. Legend may assist, as is
shown in a later chapter (p. 120). One thing is clear,
the alluvial flat of Mount's Bay, under which the submerged forest lies, formerly extended much further
seaward; and old writers mention the tradition that
St Michael's Mount formerly rose as an isolated rock
in a wood. As far as can be calculated from its
known rate of encroachment, the sea cannot have

reached the Mount till long after the Roman period, and the legend is probably quite accurate. The Mount was surrounded by a wide marshy flat covered with alders and willows till well within the historic period; the contradictory story, that the Phoenician traded to St Michael's Mount for tin seems to be the invention of a sixteenth-century antiquary.

In Mount's Bay there has been subsidence as well as loss of land through the attacks of the sea, for beneath the alluvial plain, part of which is still seen in Marazion Marsh, is buried a submerged forest. Stumps of large oaks, as well as roots of hazel and sallow, are to be seen at various points on the foreshore, where the overlying alluvium and peat have been cleared away by the sea. But the oak-stumps seem to be rooted on a soil resting directly on solid rock; they do not appear to be underlain by estuarine deposits, or by lower submerged forests. This particular land-surface may therefore represent a long period of gradual sinking, during which the trees flourished continuously, and first at a considerable elevation above the sea.

The deposit would repay closer examination, for it was not well exposed while I was staying in Cornwall. I could find no trace of man in it at Penzance, and the contained flora was principally noticeable for its poverty and the entire absence of any of the characteristic west-country plants. The trees were the oak, hazel, and sallow, the seeds obtained belong to the lesser spearwort, blackberry, a potentil, self-heal, and some sedges.

Carne, however, in 1846 was more successful at the eastern end of the Bay, for he has handed down to us an account of the strata met with in a mine-shaft on Marazion Marsh. The height of the ground at this spot is only about 12 feet above mean-tide level, and as the deposits penetrated are 32 feet thick, it is clear that both the rocky floor and the lower peat must lie beneath the level of the lowest spring tide. The position of the shaft was close to the Marazion River, where we would expect also to find an ancient buried channel. The upper deposits may be of very modern date. Commencing at the top the succession met with was :—

		Feet
	Slime, gravel and loose ground ...	8
Recent estua-rine deposits	{ Rather soft peat	4
	White sand with cockles	12
Recent or Neolithic	Layer of trees, principally oak and hazel, all prostrate. One piece of oak, about 14 feet long, appears to have been wrought, as if it had been intended for the keel of a boat ...	1 to 2
"Submerged forest"	{ Hard solid peat, of closer texture than the upper bed	3
	Alluvial gravel with tin-ore	4
	Slaty floor	at 32

It will be observed that the supposed keel of a boat occurs above the old land-surface, among drift-wood which probably belongs to the first infilling of the estuary after the submergence took place. The upper peat is probably nothing but the surface of the modern marsh, smothered and much compressed by the eight feet of "loose ground" or refuse from the neighbouring mines which had accumulated above it. The cockles probably flourished at the same level (about low-water mark) as that at which they are now found.

It is not our intention here to deal in any detail with the submerged land-surfaces noticed on the French coast opposite. The Channel Islands yield indications of submergence, and if its amount was as great as that proved on the north shores of the Channel, then the Channel Islands must have been connected with the mainland up to a period when the climatic conditions were similar to and the fauna and flora resembled those of the adjoining parts of France at the present day.

Further west, recent discoveries on the shores of the Bay of Biscay are of considerable interest, for submerged forests occur at various places, though the maximum amount of the submergence has not yet been satisfactorily made out.

One of the most interesting of the submerged forests seen between tide-marks on the French coast

was that discovered a few years since by Monsieur
Emil Gadeceau in Belle Ile. This island lies off the
mouth of the Loire, and its position some way from
the coast and well out in the Atlantic induced him
to make a special study of its flora. While engaged
in this, his attention was drawn to certain hard peaty
deposits seen only at low tide, and he asked me to
undertake the examination of the seeds found in
them. This work was gladly undertaken, as it carried
further south the examination which was then being
made into the flora of the submerged forests.

The results were somewhat surprising; out of
about 30 species sufficiently well preserved for
identification, six were no longer living in Belle Ile,
though known in Western France. The whole flora
might have come from the north of England, characteristic French species being entirely missing, though
this element is fairly represented in the living flora
of the island. In short, the flora is a northern one,
though in no degree arctic, and in this it agrees well
with the poor assemblage commonly found in the
submerged forests of the south of England.

From still further south, at various points on the
shores of the Bay of Biscay, and from the submerged
peaty deposits which underlie the Landes, seeds
have since been collected by my friend, Professor
Jules Welsch, of Poitiers. These also all belong
to common living British plants, except that at

Brétignolles, south of latitude 47°, we meet for the first time one characteristic southern plant—the vine. Unfortunately the search for traces of man and his works in these deposits has so far been unsuccessful, and we cannot yet be certain therefore that they are all of quite the same date, or correspond exactly with the submerged forests of Britain.

CHAPTER VIII

SUMMARY

To what conclusions do the foregoing somewhat monotonous pages lead? Do they help us to explain the origin of our fauna and flora? What light do they throw on the antiquity of man in Britain, or on the race problems that everywhere confront us? Can the deposits therein described be in any way connected with written history or with legend? Do they give us any approach to a measure of geological time? And, to what extent does the period of the submerged forests tie on historical times with the Glacial Epoch?

All these questions are connected with the subject-matter of this little book; but it is not written with the idea of showing how much we know or pretend to know. Our main object is to draw attention to a

much neglected period in geological history and to
suggest directions in which further research is likely
to be profitable. We have, however, made out several
points, and can give an approximate answer to some
of the questions.

It is quite clear that at the opening of the period
with which this volume deals, the greater part of
England stood fully 70 feet above its present level,
for the oldest deposit we deal with is a land-surface
covered with oak-forest and lying 60 feet below tide-
level. The oaks cannot have flourished lower, but
they may have grown on a soil well above sea-level.
Perhaps taking the whole of the evidence into account,
a subsidence of nearly 90 feet is the most probable
measure of the extent of the subsequent movement.

We do not yet know whether in England this
movement was a depression of the land or a rise
of the sea; but the fact that the relative levels
seem to have been quite different in Scotland and in
Scandinavia seems to indicate that it was the land
that moved, not the sea.

We begin, therefore, with a period when the whole
of the southern part of the North Sea was an alluvial
flat connecting Britain with Holland and Denmark,
and to some extent with France. The Isle of Wight
was connected with Hampshire, and the Channel
Islands with France. Probably the Isles of Scilly
were islands even then, for the channel between them

and Cornwall is both deep and wide, though this may possibly be due to tidal scour.

The animals and plants yet known from this lowest submerged forest are disappointingly few ; but the prevalence of the oak shows that the climate was mild, and that we have no clear indication of conditions approaching to those of the Glacial Epoch. In fact, in all the submerged forests the fauna and flora seem poor and monotonous, consisting essentially of living British species, with a few mammals since locally exterminated by man, and all known to have a wide range both in climate and latitude.

This in itself, however, is a point gained in the study of the origin of our flora; for though the deficiency is no doubt largely due to insufficient collecting, I am convinced that it is a true characteristic of this period of transition. Much time has been spent in examining and collecting the fossils of these submerged forests, and various friends have also worked at them; but everywhere we seem to get the same result, and many abandon the study because there is so little to show for it. The deposits certainly contain a much poorer fauna and flora than either the Pleistocene or the recent alluvial strata.

If we consider the Britain of the submerged forests as having lately emerged from a time when the climate was ungenial, we should naturally expect to find among the first incomers after the

change only such animals and plants as have a wide
climatic range or can migrate freely. It is these
species, and these only, which will be living on the
neighbouring lands; it is only an assemblage like
this that can stand the climatic alternations and
relapses that are likely to attend the transition.
An assemblage consisting only of species widely
distributed in latitude is probably an assemblage
that has special means of dispersal—even if we do
not happen yet to have discovered these means.

These considerations should lead us to expect to
find living, in any country which has lately undergone
a change of climate, a somewhat peculiar assemblage,
consisting mainly of hardy forms of wide range in
latitude, and not characteristically either northern or
southern. Mingled with them, we might expect a few
survivors from the previous warm or cold period. A
hardy fauna and flora seem to characterise the period
of the submerged forests; but the absence or great
scarcity of characteristic survivors from a former
period suggests that even the lowest of these deposits is far removed from the Glacial Epoch. The
arctic species had already had time to die out, or
had been crowded out; but the time had not been
sufficiently long for the incoming of the southern
forms which now characterise our southern counties.
Then, even less than now, had we reached a perfect
adjustment of the fauna and flora to the climatic

conditions; this can only be brought about by a constant invasion of species from all the surrounding regions. Some hold their own, most cannot; but as time goes on, the surviving assemblage consists more and more of species which have been able to fight against the severe competition and colonize a new country.

Garden experiments are of little use as tests of the capability of any plant to survive in this country; the study of cornfield weeds is no better. In both cases the cultivation of the land produces a bare place on which a foreign introduction has as good a chance as a native. But could this foreigner survive if the seed were dropped on a natural moor or meadow? In this connexion it is noticeable that great part of the rare British plants occur close to the coast, opposite the part of the continent in which they are found, though they are not maritime species. This is probably due to two different causes, both acting in the same direction. In the first place most of these local plants are obviously late comers, which have not yet had time to spread inland or far. And, secondly, on the coast alone do we find any considerable extent of natural bare land—practically garden land—which does not at the same time consist of poor soil. The tumbled undercliffs of our coast are just the places to give a foreign invader a chance; there only will it find patches of bare good soil, full of small cracks in which a seed is hidden from birds.

If the view is correct, that a continuous growth of our flora, and to some extent of our fauna, takes place through transportation to our coasts, from which such species as can fight their way tend more slowly to spread inland, it seems to account for the present curious distribution of species, and this in a way that no continuous land-connexion will do.

As we have pointed out in a former chapter, the land-connexion across the North Sea was a wide alluvial plain and swampy delta. What use could dry-soil plants make of such a bridge? It would be no easier for them to cross than so much sea; and migrating mammals could not greatly help in the dispersal, where so many rivers had to be crossed. The aquatic species would be helped by such a connexion, and it is curious to note that several of our most interesting aquatic plants are confined to the eastern counties, which in post-glacial times had direct connexion with the delta of the Rhine, and probably with the Elbe.

Aquatic species, however, are not dependent on continuous waterways for their dispersal; they have great facilities for overleaping barriers and reaching isolated river-basins and lakes. Every dew-pond on the downs after a few years' existence contains aquatic plants and mollusca, and a still larger number of species, including fish, will be found in ancient flooded quarries or prehistoric dykes surrounding some hill-fortification. If an aquatic plant is fairly common on the continent near by, it is almost certain to occur in some isolated pond or river in the part of Britain opposite.

Many of our peculiar mollusca and plants are limestone species, which must have crossed over at a single leap, for no elevation or depression will connect the various isolated limestone masses of Britain. A post-glacial elevation would connect the North Downs with the corresponding chalk-hills of France; but these Downs are isolated by wide tracts of non-calcareous strata from the areas of Oolite or Carboniferous limestone to which many of our limestone animals and plants are now confined. There is also nothing in the present distribution of our limestone species to suggest that any great stream of migrants used this bridge of chalk-downs.

It may be asked, Why discuss these questions here, if all these peculiar species are unknown in the submerged forests? In certain cases negative evidence is of great value, and the deficient flora of the submerged forests is a case in point. We find a striking contrast between this ancient flora and the flora which flourished when cultivation of the land had begun. The Roman deposits in Britain yield many species which have not yet been found in the submerged forests, and even the earlier Celtic deposits have already yielded a few of them. To a large extent this difference is due to the agency of man, intentional to a certain extent, but mainly accidental, through the introduction of weeds and the preparation of the soil for crops. It must not be forgotten that man not only introduced the weeds, he prepared the land on which they could establish themselves, and from thence spread to uncultivated ground where few botanists now suspect that they are anything but "native."

In days when the people of Britain were hunters, the only extensive open country in the south and east seems to have been the chalk-downs and the sandy heaths. These were not suitable for new additions to the plant population, for the good land was all oak forest, the barren heaths were unfavourable for any but heath plants, and the alluvial flats were largely covered with sallow and alder. The open downs were clothed with close turf, and until this was broken by cultivation there would be little chance for migrants. It seems, therefore, that to obtain a clear idea of the plant population of this country before man's influence could be felt, we must study the flora of the submerged forests and of the associated alluvial detritus washed from the uplands during the same period. Till this is done more thoroughly, it is not much use to discuss what species are "native" and what "introduced"; the submerged forest will yield the answer to this question.

The next question we have put—What light do these submerged forests throw on the antiquity of man in Britain, or on the race-problems of Britain?— is a difficult one to answer in the present state of our knowledge. Valuable evidence has been lost through the failure to preserve most of the human remains that have been found; but both Owen and Huxley recognised the peculiar type of the "river-drift man." Unfortunately few implements have been collected, and the pieces of wood shaped by man, though recorded, have not been preserved. One implement of polished stone has certainly been found in the latest submerged land-surface, but it is not clear that any-thing except flakes has been obtained in the older deposits. Still the stratigraphical relations seem to indicate that all these deposits are of Neolithic age and later than the Palaeolithic terraces. The re-lations of Palaeolithic to Neolithic are still very obscure in this country, and the reason is perhaps to be sought in a submergence which has tended to carry many of the transition deposits beneath the sea-level, or has caused them to be silted up under more modern alluvium. The lowest submerged forest requires careful search before we can be certain of its true position in the sequence; but it is seldom exposed, and then only in dock-excavations soon again hidden.

Before we can attempt to answer the other questions, it is important to get an estimate of the

amount of time occupied in the formation of these
deposits, and of the lapse of time since the last
of them was formed. The newest of them belongs
certainly to the age of polished stone, and the earliest
also probably comes within the Neolithic period. We
have already seen that within the period represented
by the submerged forests there has been a rise of the
sea-level, or depression of the land, to the extent of
80 feet, perhaps a few feet more. If we can obtain
some measure of the time occupied in the formation
of such a series of deposits, this should give us some
idea as to the length of the Neolithic period, and
also of the rate at which changes of the sea-level
sometimes can take place.

It is unfortunate that for these calculations so
many of the factors are of uncertain value. We
may estimate from the present rate of erosion of
the coast the amount that has been lost since the
sea-level became stationary, or we may take the rate
of accumulation of sand-dunes or shingle-spits; or
the rate at which our estuaries, harbours, and broads
are silting up. It all comes, however, to this—no
exact figures can be given; but so many rough calculations lead to approximately the same date, that
the date arrived at may be trusted to give some idea
of the length of the period which has elapsed since
the downward movement ceased.

Working backwards from the present day, step by

step, archaeological evidence gives an undoubted
period of 2000 years, to the first century B.C., during
which no measurable change of sea-level has taken
place in the south of England.

To this must be added a few centuries for the
growth of the marshes on which Glastonbury and
similar lake-dwellings were built, and for the growth
of various other marshes at present sea-level known
to be earlier than the Roman invasion.   Also we
must allow for the accumulation of various shingle-
spits and sand-dunes then already partly formed.

In general, somewhere about one-third or one-
half of this accumulation seems to have taken place
before the Roman invasion.   This adds another 1500
years ; so that about 3500 years ago, we get back to
the beginning of the period of unchanging sea-level
in which we are still living, and begin to see evidence
of earth movements still in progress.

Whether this 3500 years will take us back to the
beginning of the Bronze Age in Britain is not yet
proved; but so far we seem to discover metals in the
whole of the deposits formed whilst the sea-level remained unchanged, and only stone weapons in even
the newest of the submerged forests.   For the present,
we may therefore take it that the two changes nearly
coincided.   The use of metals began in Britain about
the time that the earth-movements ceased—that
is to say somewhere about 1600 B.C.

Whether this period of 3500 years will really take us back to the commencement of the Bronze Age is doubtful, for Stonehenge had already been built, and though only stone hammers seem to have been there used, yet one slight streak of bronze or copper has been noticed. Of course, there may have been a similar occasional use of bronze at the time of the last submerged forest; but we have as yet no evidence of this, and the possible correspondence in date between Stonehenge and the last of the submerged forest remains merely a suggestion.

Perhaps we may still find submerged stone-circles or other antiquities of the age of Stonehenge beneath the sea-level; but Stonehenge lies too high above the sea for it in itself to give any clue as to a change of sea-level. We will only make one suggestion. It is probable that when Stonehenge was built, a long arm of the sea extended far up the Avon Valley, so that navigable water was found not far from Stonehenge. There is in Stonehenge an inner circle of smaller stones, not composed of the local greywethers but consisting of large blocks of igneous rock of foreign origin. These blocks, which are sufficiently large to be awkward for land-carriage, have been said to be erratics gathered on Salisbury Plain, just as the grey-wethers for the main circle were gathered; but there are no erratics on Salisbury Plain. Large erratic blocks of similar character occur, however, abundantly on the lowlands of Selsey Bill, under the lee of the Isle of Wight. Probably a similar erratic-strewn plain once fringed the coast on the west also, though on the exposed side the part above the sea-level has now been entirely swept away by the sea.

I would suggest that the Stonehenge erratics, instead of being brought from any great distance, may have come from a wide plain at the mouth of the Avon, then two or three miles further seaward. From thence they were rafted far up the navigable fjord, not yet silted up, and were only carried a short distance uphill. Igneous rocks such as these, found in a country consisting essentially of chalk and Tertiary strata, would be valuable and probably endowed with magic properties, hence their employment in this inner circle.

Our next enquiry must be into the length of time represented by the series of submerged forests and associated deposits described in the foregoing pages. The newest of them belongs certainly to the age of polished stone, and the earliest also probably comes within the Neolithic Period. Within the period represented by the submerged forests, we have seen that there has been a change of the sea-level to the extent of 80 feet, or perhaps rather more. If we can obtain some measure of the time occupied, this should give us some approximate idea as to the length of

the Neolithic period, and of the rate at which changes
of the sea-level can take place.

The first point to be considered is the length
of time occupied by the growth of the series of
submerged forests. On first examining, or reading
accounts of, deposits of this sort one obtains a vague
impression of long periods, during which mighty oaks
flourished. Both the movements of submergence and
the intervening periods of vegetable growth seem to
require great lapses of time. On closer study, however, the evidence seems scarcely to support this
view, for estuarine silts are deposits of exceptionally
rapid growth, and one finds that the usual characteristic of a "submerged forest" is that it shows indications of only a single generation of trees. The trees
also are usually small, except where the submerged
forest rests directly on deposits of much earlier date,
or on solid rock.

It should be remembered that the large oak trees
which are often found in the lowest land-surface at
any particular place do not necessarily belong to any
one special stage of the submergence. These same
trees may have grown continuously above tide-
marks during several successive stages, until at
last the upward creeping water rose sufficiently
to reach this part of the forest. The large well-
grown oaks seen in Mount's Bay and various other
places are, as far as I have seen, all rooted on

ancient gravels, solid rock, or boulder clay, not on beds of silt.

We cannot speak confidently as to the time needed to form each thin layer of vegetable soil, marsh peat, or estuarine silt. On comparing the submerged land-surfaces, however, with similar accumulations formed within known periods, such as marsh soils grown behind ancient embankments, or forest-growth over flats silted up at known dates, we can learn something. No one of the land-surfaces alternating with the silts would necessarily require more than a century or two for its formation. Brushwood and swamp growth are the characteristic features of these deposits, and such growth accumulates and decays very rapidly. Possibly trees of older growth may still be found, but I have not succeeded in discovering a tree more than a century old in any one of the marsh deposits alternating with the estuarine silts. Oaks of three centuries may be observed rooted in the older deposits; but this, as above explained, is another matter.

It is useless to pretend to any exact calculations as to the time needed for the formation of these alternating strata of estuarine silt and marsh-soil; but looking at the whole of the evidence without bias either way, it seems that an allowance of 1000, or at most 1500, years would be ample time to allow. A period of 1500 years may therefore be taken to cover the whole of the changes which took place during the period of gradual submergence.

If this is approximately correct, the date at which the submergence began was only 5000 years ago, or about 3000 B.C. The estimate may have to be modified as we obtain better evidence; but it is as well to realize clearly that we are not dealing with a long period, of great geological antiquity; we are dealing with times when the Egyptian, Babylonian, and Minoan civilizations flourished. Northern Europe was then probably barbarous, and metals had not come into use; but the amber trade of the Baltic was probably in full swing. Rumours of any great disaster, such as the submergence of thousands of square miles and the displacement of large populations might spread far and wide along the trade routes. Is it possible that thus originated some of the stories of the deluge?

We will not now pursue this enquiry; but it is well to bear in mind the probability that here geology, archaeology, and history meet and overlap. Any day one of our submerged forests may yield some article of Egyptian manufacture of known date, such as a scarab, which has passed from hand to hand along the ancient trade routes, till it reached a country still living in the Stone Age, where its only use would be in magic. But it might now serve to give us a definite date for one of these submerged forests. It might happen to have been lost with some of the stone implements, or with one of the human skeletons, apparently belonging to persons drowned, for no trace of a grave is ever mentioned. A find of this sort is no more improbable than the discovery of a useless modern revolver in a bag of stone and bone tools belonging to some Esquimaux far beyond the reach of ordinary civilized races.

In this connexion it might be worth while systematically to dredge the Dogger Bank, in order to see whether any implements made by man can be found there. The alluvial deposits are there so free from stones that if any at all are found in them they may probably show human workmanship. The Dogger Bank may have remained an island long after great part of the bed of the North Sea had been submerged, for the Bank now forms a submerged plateau. It may even have lasted into fairly recent times, the final destruction of the island being due to the planing away of the upper part of the soft alluvial strata through the attacks of the sea and of boring molluscs. *Pholas* is now actively attacking the hard peat-beds at a depth of more than 10 fathoms, and is rapidly destroying this accumulation of moorlog, wherever the tidal scour is sufficient to lay it bare.

BIBLIOGRAPHY

BRÖGGER, W. C. Om de Senglaciale og Postglaciale Nivåforandringer i Kristianiafeltet (Molluskfaunan). *Norges geol. undersögelse.* 1900–1.

COLENSO, J. W. A Description of Happy-Union Tin Stream-Work at Pentuan. *Trans. R. Geol. Soc. Cornwall,* vol. IV, 1832, p. 30.

[GODWIN-]AUSTEN, R. A. C. On the Valley of the English Channel. *Quart. Journ. Geol. Soc.* vol. VI, 1849, p. 69.

GODWIN-AUSTEN, R. A. C. On the Newer Tertiary Deposits of the Sussex Coast. *Quart. Journ. Geol. Soc.* vol. XIII, 1856, p. 40.

HENWOOD, W. J. On some of the Deposits of Stream Tin-Ore in Cornwall, with remarks on the Theory of that Formation. *Trans. R. Geol. Soc. Cornwall,* vol. IV, 1832, p. 57.

HOLMES, T. V. Excursion to Tilbury Docks, May 17, 1884. Record of Excursion, *Geol. Assoc.* p. 182.

—— Notes on the Geological Position of the Human Skeleton lately found at the Tilbury Docks, Essex. *Trans. Essex Field Club,* vol. IV, 1884, p. 135.

KEITH, Prof. A. Ancient Types of Man. London, 1911.

LORIÉ, Dr J. Contributions à la Géologie des Pays-Bas. V. Les Dunes intérieures, les Tourbières basses et les Oscillations du Sol. *Archives Teyler,* Sér. II, t. III, 5ᵐᵉ Partie. 1890.

MILLER, S. H., and S. B. J. SKERTCHLY. The Fenland past and present. 8vo. London, 1878.

MORTON, G. H. Local Historical, Post-glacial and Pre-glacial Geology. *Proc. Liverpool Geol. Soc.* 1887–8.

MUNTHE, H. Studier öfver Gottlands Senkvartära Historia. *Sveriges Geol. Undersökning.* 1910.

OWEN, Sir R._ Antiquity of Man. *Proc. Roy. Soc.* vol. xxxvi, 1883, p. 136 ; and more fully in his *Antiquity of Man as deduced from the Discovery of a Human Skeleton at Tilbury.* 8vo. London, 1884.

POTTER, C. Observations on the Geology and Archaeology of the Cheshire Shore. *Trans. Hist. Soc. Lancashire and Cheshire.* 1876.

PRESTWICH, Sir J. Geology. Vol. ii. Oxford, 1888.

PREVOST, Dr E. W. and others. The Peat and Forest Bed at Westbury-on-Severn. *Proc. Cotteswold Field Club*, vol. xiv, 1901, p. 15.

RASHLEIGH, P. An Account of the Alluvial Depositions at Sandry-cock. *Trans. R. Geol. Soc. Cornwall*, vol. ii, 1822, p. 281.

READE, T. M. The Geology and Physics of the Post-glacial Period, as shown in the Deposits and Organic Remains in Lancashire and Cheshire. *Proc. Liverpool Geol. Soc.* vol. ii, session xiii, 1871–2, p. 36.

——— Glacial and Post-glacial features of the Lower Valley of the River Lune and its estuary. *Proc. Liverpool Geol. Soc.* vol. ix, session xliii, 1902, p. 163.

REID, C. On the Relation of the present Plant Population of the British Isles to the Glacial Epoch. *Rep. Brit. Assoc. for* 1911, p. 573.

——— The Island of Ictis. *Archaeologia*, vol. lix, 1905.

——— On the former Connexion of the Isle of Wight with the Mainland. *Rep. Brit. Assoc. for* 1911, p. 384.

Reid, C. The Geology of the country around Southampton. *Mem. Geological Survey*, 1902.

—— The Geology of Holderness. *Mem. Geological Survey.*

—— The Origin of the British Flora. 8vo. London, 1899.

Rogers, Inkermann. On the Submerged Forest at Westward Ho! Bideford Bay. *Trans. Devon. Assoc.* vol. xl, 1908, p. 249.

Shore, T. W., and J. W. Elwes. The New Dock Excavation at Southampton. *Proc. Hampshire Field Club*, 1889, p. 43.

Shore, T. W. Hampshire Mudlands and other Alluviums. *Proc. Hampshire Field Club*, 1893, p. 181.

Skertchly, S. B. J. Geology of the Fenland. *Mem. Geological Survey*, 1877.

Spurrell, F. C. J. Early Sites and Embankments on the Margins of the Thames Estuary. *Archaeological Journ.* 1885.

—— On the Estuary of the Thames and its Alluvium. *Proc. Geol. Assoc.* vol. xi, 1889, p. 210.

Stather, J. W. Shelly Clay Dredged from the Dogger Bank. *Quart. Journ. Geol. Soc.* vol. lxviii, 1912, p. 324.

Strahan, A. On submerged Land-surfaces at Barry, Glamorganshire. With notes on the Fauna and Flora by Clement Reid. *Quart. Journ. Geol. Soc.* vol. lii, 1896, pp. 474—489.

Tylor, A. On Quaternary Gravels. *Quart. Journ. Geol. Soc.* vol. xxv, 1869, p. 57.

Whitaker, W. The Geology of London and of part of the Thames Valley. *Mem. Geol. Survey*, 1889, 2 vols.

Whitehead, H., and H. H. Goodchild. Some Notes on "Moorlog," a peaty deposit from the Dogger Bank in the North Sea. *Essex Naturalist*, vol. xvi, 1909, p. 51.

Worth, R. H. Evidences of Glaciation in Devonshire. *Trans. Devon. Assoc.* vol. xxx, 1898, p. 378.

INDEX

Printed in Great Britain
by Amazon.co.uk, Ltd.,
Marston Gate.